Sandford Fleming

Report on the Intercolonial Railway Exploratory Survey

Made under Instructions from the Canadian Government in the Year 1864

Sandford Fleming

Report on the Intercolonial Railway Exploratory Survey
Made under Instructions from the Canadian Government in the Year 1864

ISBN/EAN: 9783337216191

Printed in Europe, USA, Canada, Australia, Japan

Cover: Foto ©berggeist007 / pixelio.de

More available books at **www.hansebooks.com**

REPORT

ON THE

INTERCOLONIAL RAILWAY

EXPLORATORY SURVEY,

UNDER INSTRUCTIONS FROM THE CANADIAN GOVERNMENT, IN THE YEAR 1864.

BY SANDFORD FLEMING,
CIVIL ENGINEER.

Printed by Order of the Legislative Assembly.

Quebec:
HUNTER, ROSE & CO., 26, ST. URSULE STREET.
1865.

CONTENTS.

	PAGE.
Letter presenting Report.	1
Schedule of Plans and Profiles submitted.	2 & 3
Instructions and Preliminary Correspondence.	4
Report.	8
The Engineering staff.	8
Main Divisions of the survey.	9
The Nova Scotia Division of the survey.	9
Line No. 1.	10
Line No. 2.	10
Line No. 3.	10
Line No. 4.	11
Tables of Gradients.	11
Distances by the several Lines.	12
Lines Nos. 5 & 6.	12
Estimate of quantities.	13
New Brunswick and Canada Division of the Survey.	15
Air Lines.	15
The Surveyed Central Line.	19
River du Loup to River Trois Pistoles.	19
River Trois Pistoles to Green River Forks.	20
Green River Forks to R. Restigouche.	22
R. Restigouche to R. Tobique.	23
River Tobique to Keswick Summit.	24
Keswick Summit to Little River.	26
Little River to Coal Creek.	27
Coal Creek to Apohaqui Station.	28
Character of Grades on whole Line.	29
Approximate quantities, ditto.	31
The Matapedia Survey (70 mile section).	31
Character of grades.	34
Curvature.	35
Approximate quantities.	35
Datum Levels.	35
Fitness for settlement.	36
Central District.	36
Matapedia District.	37
Various projected Routes.	38
Frontier Routes.	38
Line No. 1.	38
Line No. 2.	39
Line No. 3.	40
Central Routes.	41
Line No. 4.	41
Line No. 5.	41
Line No. 6.	42
Line No. 7.	42
Line No. 8.	42
Line No. 9.	43
Line No. 10.	44
Line No. 11.	44
Line No. 12.	45
Bay Chaleur Routes.	45
Line No. 13.	45
Line No. 14.	46
Line No. 15.	47
Comparative Distance from River du Loup, to St. John and Halifax.	48
Distance of the several lines from the U. S. Frontier.	50

		PAGE.
Commercial advantages of different Routes............		50
Local Traffic.........		51
Through Freight Traffic............		52
Through Passenger Traffic.........		53
Great Shippigan Harbour..........		54
Distance to Liverpool, England.........		54
do	to Quebec.......	55
do	to Montreal........	55
do	to Toronto......	55
do	to Buffalo.........	55
do	to Detroit	55
do	to Chicago........	55
do	to Albany........	55
do	to New York.......	55
do	to St. John, N. B........	56
Climatic Difficulties.........		57
Effects of Frost........		58
Heavy snow-falls..........		58
The Estimate of probable cost.........		60
Remarks on Engineering, (Expenditure on).........		60
do	Right of way and fencing, do	61
do	Clearing, do	61
do	Dwellings for workmen, do	62
do	A Telegraph, do	62
do	Bridging and Grading, do	62
do	Superstructure, do	63
do	Station Accommodation, do	64
do	Rolling Stock, do	64
do	Contingencies.........	64
do	·Uniform mileage charges	65
Estimates—Nova Scotia Division of Survey........		65
do	Surveyed Line River du Loup to Apohaqui.......	65
do	Bay Chaleur Line........	66

APPENDICES.

(A)—The Agricultural capabilities of New Brunswick—
    As indicated by its Geological structure................. 67
        " by a practical survey and examination of its soils..... 70
        " by actual and comparative productiveness,—Professor Johnston......... 75
(B)—Agricultural capabilities of the Matapedia district,—A. W. Sims............. 80
(C)—Frontier Route, Line No. 1. Report on an examination of the country between River du Loup and Woodstock,—T. S. Rubidge, C. E..................... 83
(D)—Frontier Route, Line No. 2. Correspondence in reference to the extension of the St. Andrews and Woodstock Railway,—Walter M. Buck, C. E............. 88
(E)—Central Route, Line No. 8. Report on Exploration from Boiestown across the Tobique Highlands,—W. H. Tremaine, C. E............. 93
(F)—Remarks on the shortest lines of communication between America and Europe in connection with the contemplated Intercolonial Railway 95
Maps printed to accompany Report—
    General Map of the country between Quebec and Halifax, showing the various projected routes.
    Chart showing the relative geographical position of the British Islands and British America with the shortest lines across the Atlantic, to accompany Appendix F.

REPORT

ON THE

INTERCOLONIAL RAILWAY SURVEY.

MONTREAL, February 9th, 1865.

To the Honorable WILLIAM MCDOUGALL,
Provincial Secretary, Canada.

SIR,—I have the honor to submit the following Report on the Explanatory Survey of the Territory through which the contemplated Railway between the Provinces of Canada, New Brunswick and Nova Scotia is intended to run.

In conducting this Survey, I have considered the routes for the projected Railway which have, on previous occasions, been contemplated, as well as some others which seemed worthy of attention.

I have especially directed my attention to the best means of overcoming or avoiding obstacles which were previously considered serious or insuperable.

I have endeavored to carry on the survey with a strict regard to economy, at the same time efficiency, and I have completed the whole service at as early a period as it was possible, with the means at my command.

I shall, in the following pages, describe the quality of the land in the country examined, and its fitness for cultivation and settlement so far as I have been able to acquire information. I shall also make some allusion to the climatic influences which may operate on the several routes.

I shall likewise report, although I fear imperfectly, on the comparative advantages of the various routes, in a commercial point of view.

The relative position of the several projected routes with the frontier of the United States, will be described.

The estimates of probable cost will be based on calculations made with a view to efficiency, stability and permanency; at the same time having due regard to economy in the expenditure.

A schedule of the plans and profiles of the several lines surveyed, and explorations made, and which have been laid down to convenient scales; together with other papers relating to the survey, will be found subjoined.

I trust that the information which I have now the honor to submit will enable the Government to judge of the practicability, probable cost, and respective merits, of the several projected routes of this proposed intercolonial communication.

The Governments of the Sister Provinces have afforded me every facility in the prosecution of the Survey, and I am under no ordinary obligations to many of the leading gentlemen in New Brunswick and Nova Scotia for their ready assistance and the valuable information with which they have furnished me.

I have the honor to be, Sir,
Your obedient servant,
SANDFORD FLEMING.

SCHEDULE OF PLANS AND PROFILES SUBMITTED.

1. Plan of Surveyed Line from Trois Pistoles to Snellier River. Length of Line, 38 miles. Scale, 500 feet to one inch.
2. Approximate Profile of Line from Trois Pistoles to River Snellier. Scales, Horizontal 500 feet, Vertical 50 feet to one inch.
3. Plan of Surveyed Line from Snellier River to Green River Forks. Length of Line, 45 miles. Scale, 500 feet to one inch.
4. Approximate Profile of Line from Snellier River to Green River Forks. Scales, Horizontal 500 feet, Vertical 50 feet to one inch.
5. Plan of Surveyed Line from Green River Forks to Restigouche. Length of Line, 34 miles. Scale, 500 feet to one inch.
6. Approximate Profile of Line from Green River Forks to Restigouche. Scales, Horizontal 500 feet, Vertical 50 feet to one inch.
7. Plan of Surveyed Line from Restigouche to Tobique. Length of Line, 45 miles. Scale, 500 feet to one inch.
8. Approximate Profile of Line from Restigouche to Tobique. Scales, Horizontal 500 feet, Vertical 50 feet to one inch.
9. Plan of Surveyed Line from Tobique to Miramichi Forks. Length of Line, 37 miles. Scale, 500 feet to one inch.
10. Approximate Profile of Line from Tobique to Maramichi Forks. Scales, Horizontal 500 feet, Vertical 50 feet to one inch.
11. Plan of Surveyed Line from Miramichi Forks to Keswick Summit. Length of Line, 55 miles. Scale, 500 feet to one inch.
12. Approximate Profile of Line from Miramichi Forks to Keswick Summit. Scales, Horizontal 500 feet, Vertical 50 feet to one inch.
13. Plan of Surveyed Line from Keswick Summit to Little River. Length of Line, 61 miles. Scale, 500 feet to one inch.
14. Approximate Profile of Line from Keswick Summit to Little River. Scales, Horizontal 500 feet, Vertical 50 feet to one inch.
15. Plan of Surveyed Line from Little River to Coal Creek. Length of Line, 26 miles. Scale, 500 feet to one inch.
16. Approximate Profile of Line from Little River to Coal Creek. Scales, Horizontal 500 feet, Vertical 50 feet to one inch.
17. Plan of Surveyed Line from Coal Creek to Apohaqui. Length of Line, 32 miles Scale, 500 feet to one inch.
18. Approximate Profile of Line from Coal Creek to Apohaqui. Scales, Horizontal 500 feet, Vertical 500 feet to one inch.
19 Plan of Surveyed Line from Parsboro' to Truro. Length of Line, 60 miles. Scale, 500 feet to one inch.
20. Approximate Profile of Line from Parsboro' to Truro. Scales, Horizontal 500 feet, Vertical 50 feet to one inch.
21. Plan of Surveyed Line from the River Metis to Pierre Brucho's. Length of Line, 30 miles. Scale, 200 feet to one inch.
22. Approximate Profile of Line from River Metis to Pierre Brucho's. Scales, Horizontal 200 feet, Vertical 30 feet to one inch.
23. Plan of Surveyed Line from Pierre Brocho's on Lake Matapedia to near the Forks. Length, 30 miles. Scale, 200 feet to one inch.
24. Approximate Profile of Line from Pierre Brucho's on Lake Matapedia to near the Forks. Scales, Horizontal 200 feet, Vertical 30 feet to one inch.
25. Plan of Surveyed Line from third mile below the Forks of the Matapedia to the Restigouche. Length of Line, 32 miles. Scale 200 feet to one inch.

26. Approximate Profile of Line from the third mile below the Forks of the Matapedia to the Restigouche. Scales, Horizontal 200 feet, Vertical 30 feet to one inch.

27. Profile of Line Surveyed from near Moncton to Tautramar Marsh, near Sackville, by Mr. Boyd, distance 30 miles. Scales, Horizontal 400 feet, Vertical 60. feet to an inch.

28. Plan of Exploration for alternative line between Rivers Restigouche and Tobique. Scale, one mile to an inch.

29. Plan of Explorations in the Highland District at the Sources of the Rivers Rimouski, Kedgwich, Green River, Snellier, Turadi, and Toledi, with Barometrical elevations. Scale, one mile to an inch.

30. General Map of the Country between Quebec and Halifax, showing the Lines Surveyed and Projected. Scale, 8 miles to an inch.

31. Chart shewing the Relative Geographical Position of the British Islands and British America, with the Shortest Great Lines of Communication between the Continents of Europe and America.

32. Plan of the Line Surveyed in 1864, from St. John, N.B., to Fredericton, by Mr. Burpee. Length, 65 miles.

33 Profile of the Line Surveyed in 1864, from St. John to Fredericton, and to St. Andrew's Junction, by Mr. Burpee.

34. Approximate Profile of Line from River du Loup to River Trois Pistoles from Mr. Rubidge's Survey, 1858. Length, 24½ miles. Scales, Horizontal 400 feet, Vertical 40 feet to an inch.

35. Plan of Line by Acadia Mines from Truro to Rufus Black's on River Phillip. Length, 41 miles. Scale, 5 chains to an inch. Mr. Beattie's Survey, 1864.

36. Profile of Line by Acadia Mines. Length, 41 miles. Scales, Horizontal 5 chains, Vertical 50 feet to an inch.

INSTRUCTIONS

TO

SANDFORD FLEMING, C. E.,

FROM

THE HONORABLE THE PROVINCIAL SECRETARY OF CANADA.

SECRETARY'S OFFICE,
Quebec, 11th March, 1864.

SIR,—I now address to you in writing, instructions by the Government of Canada for the survey intrusted to you of the route of the proposed Intercolonial Railway, the substance of which instructions has already been communicated to you in a verbal manner, such mode of communication having been adopted at the time in order to avoid delay in your departure from Quebec on the duty in question.

1. You are instructed on the part of the Government of Canada, to proceed immediately to a survey and examination of the territory through which the proposed Railway between this Province and those of New Brunswick and Nova Scotia would run.

2. This survey and examination are intended for the purpose of enabling the Government of Canada to form an estimate of the practicability of the proposed undertaking, and of its probable cost, in order that the expediency of engaging in the work itself may be judged of in a satisfactory manner.

3. The information so obtained will also be at the service of the other Governments interested if desired.

4. On a general examination of the country, you will consider the routes which have on previous occasions been contemplated for the object in question, as well as any others which may seem to you worthy of attention.

5. Your notice will be especially given to any obstacles which may present themselves as requiring serious expense to surmount, and to the best methods of overcoming such obstacles, or of avoiding them by deviations from the direct line.

6. You will also pay attention to the distance of what may in other respects appear the most eligible line from the frontier of the United States at various points.

7. You will make your calculations in the matter of the probable cost of the work with a due regard to economy, but at the same time to full efficiency.

8. Similar considerations will guide you as regards the survey and examination.

9. You will endeavor to act in a cordial and harmonious spirit with any persons who may be appointed, either on the part of the sister colonies or of the Imperial Government, to co-operate with you.

10. The completion of the survey and examination at as early a period as possible is highly desirable.

11. You will report your progress from time to time to the Provincial Secretary of Canada.

I have the honor to be, Sir,
Your obedient servant,
(Signed,) A. J. FERGUSSON BLAIR,
Secretary.

S. FLEMING, Esquire,
Civil Engineer, Fredericton, N. B.

Letter from Sandford Fleming to the Hon. the Provincial Secretary, Canada.

[Copy.]

HALIFAX, 25th April, 1864.

The Honorable
    The Provincial Secretary, Canada.

SIR,—I had the honor, on the 21st of March last, to receive at Boiestown, in New Brunswick, written instructions, dated Quebec, 11th March, respecting the survey of the comtemplated Intercolonial Railway, which I had previously been conducting under verbal and general instructions.

By these instructions I was directed on the part of the Government of Canada to survey and examine the territory through which the proposed line of Railway between the provinces of Canada, New Brunswick and Nova Scotia would run, in order that an estimate may be formed of the practicability of the proposed undertaking, the probable cost of such line or lines as might appear most eligible and their positions in respect to the frontier of the United States. I was further directed to report progress from time to time.

I have now the honor to report that I have made a general reconnoissance of a great portion of the country between this place and the present terminus of the Grand Trunk Railway at River du Loup, that I have instituted exploratory surveys across from the St. Lawrence to the head waters of the River Restigouche, from the River Tobique to the River Miramichi near Boiestown, and from the last named place to the line of Railway now built from St. John to Shediac. These surveys are not yet sufficiently far advanced to enable me to report on the probable results.

A considerable quantity of provisions for the use of surveying parties, during the ensuing summer, has been purchased and forwarded to the interior of the country; these provisions are placed in store on the height of land between the St. Lawrence and the Restigouche, at a convenient point to farther surveying operations. I have endeavored to employ the winter season to the best advantage, and I now intend to prosecute the survey with vigor in order that it may be satisfactorily completed, agreeably to the desire expressed in my instructions, at as early a period as possible; with that object in view I am organizing a sufficient number of surveying parties to assist me in the important work with which I have been intrusted. These parties will take the field at once, and in order to defray the cost of the requisite outfit and current expenses, I will before long make a requisition for funds.

It gives me great pleasure to state that the Governments of New Brunswick and Nova Scotia have furnished me with every information in their possession, and have afforded me every facility in the prosecution of the survey so far. The latter Government has requested me to act as Railway Engineer for Nova Scotia, thus evincing a desire to act in harmony with the Canadian Government in completing the great work of Railway communication between the Provinces.

I return at once to New Brunswick, where I will be engaged for a short period, after which I shall proceed to Canada, for the purpose of completing arrangements for carrying on active operations during the summer.

I may take this opportunity of stating that any communication with which you may be pleased to honor me will soonest reach me during the progress of the survey if addressed Quebec.

    I have the honor to be, Sir,
        Your obedient servant,
    (Signed,)   SANDFORD FLEMING.

Letter from Sandford Fleming to the Honorable the Provincial Secretary, Canada.

QUEBEC, May 5th., 1864.

To the Honorable
    The Provincial Secretary, Canada.

SIR,—I had the honor to address you from Halifax, on the 25th April last, on the subject of the Intercolonial Railway Survey, reporting the progress made and indicating the steps now being taken by me to prosecute the Survey agreeably to instructions.

I have now the honor to inform you that I have this morning arrived from New Brunswick, and that I am losing no time in completing arrangements to have a sufficient number of surveying parties in the field as early as possible.

A continuous supply of funds will be required to carry on the survey as at present contemplated, of not less than $3000 per month, and it would greatly facilitate the work if I had the authority to draw to that amount through any of the Bank Agencies in the Lower Provinces, where the expenditure will chiefly take place.

This rate of expenditure during the present year will not, it is true, be sufficient to make perfect surveys and working plans, but it will, I feel somewhat confident, be sufficient to enable the Government to form an estimate of the practicability of the proposed undertaking, as well as the comparative cost of some of the routes spoken of.

The expenditure through me up to this time has been $2,900, in addition to which a further sum has been paid by the Government for the purchase of supplies and forwarding them to the interior of the country for future use. I am not aware what amount has been so expended, but it is probable that up to this time the survey has cost not less than $6000, leaving a balance of the amount appropriated last year of $4,000.

It will thus be evident from the rate of expenditure contemplated, that an additional sum of $20,000 will be required during the present year. I have respectfully to request that sufficient funds be placed at my disposal to pay the current expenses of the service which I have the honor to conduct. I will be happy to furnish at any time statements of expenses with vouchers.

I have the honor to be, Sir,
Your most obedient servant,
(Signed,) SANFORD FLEMING.

Letter from the Honorable the Provincial Secretary of Canada, to Sandford Fleming.

SECRETARY'S OFFICE,
QUEBEC, 6th May, 1864.

SIR,—I have the honor to acknowledge the receipt of your letter, dated Halifax, 25th ult., and of your second letter, dated Quebec, the 5th inst., upon various topics connected with the survey of the proposed Intercolonial Railway line.

Being fully aware that the members of the Government are extremely anxious that the survey upon which you are engaged shall be energetically prosecuted, in order that they may as speedily as possible be placed in possession of the important information expected to result from it, I shall be very glad, if you will enable me, when formally submitting these communications for the consideration of my colleagues, to lay before them at the same time your own opinion of the period at which such survey will be completed.

I have the honor to be, Sir,
Your obedient servant,
(Signed,) JOHN SIMPSON, Secretary,

S. FLEMING, Esquire.
Civil Engeneer, Quebec.

Letter from Sandford Fleming to the Honorable the Provincial Secretary, Canada.

QUEBEC, May 6th, 1864.

SIR,—I have the honor to acknowledge the receipt of your letter of this date, in which you desire me to state when in my opinion the survey of the proposed Intercolonial Railway will be completed. The instructions, dated 11th March last, which I had the honor to receive, and under which I am now acting, appear to me to mean that what may be termed a "Preliminary Exploratory Survey" is contemplated; that I should be prepared to report as soon as possible on the various routes which have been proposed, so as to give the Government a tolerably correct idea of the practicability and the cost of each, the nature of the difficulties requiring serious expense to surmount, the character of the country through which they pass, and their position with respect to the frontier of the United States.

To make this survey, I propose to direct my attention chiefly to the difficult points on each route, and more especially to that portion of the central route lying between Miramichi and the boundary of Canada ; on that portion and at the points referred to I shall make surveys of such a character as will satisfy myself as to the practicability or otherwise of the line as well as the approximate cost of overcoming obstacles of a serious nature. Where the county is comparatively level and a line easily constructed, a general examination will probably suffice.

A survey of this nature can, I think, be completed within the present year, at a cost not greatly exceeding the estimate I had the honor to submit in my communication of yesterday's date. A more exact and thorough survey, should the Government desire it, will of course require a much larger outlay.

I have the honor to be, Sir,
Your most obedient servant,
(Signed,) SANDFORD FLEMING.

The Hon. JOHN SIMPSON,
Provincial Secretary, Canada.

Letter with additional instructions from the Honorable the Provincial Secretary, Canada, to Sandford Fleming.

SECRETARY'S OFFICE,
QUEBEC, 7th May, 1864.

SIR,—I have the honor to acknowledge the receipt of your letter of yesterday's date—which, with your two previous communications on the same subject, namely, the Intercolonial Railway Survey, the Executive Council have had under their consideration.

And I am directed to request that, in addition to the subject mentioned in your letter of yesterday, as those to which in making the survey you propose to direct your chief attention, you will report as accurately and distinctly as possible upon the following topics:

1. The comparative advantages of the various routes embraced in your survey, in a commercial point of view.
2. The quality of the land on the several routes and fitness for cultivation and settlement.
3. The climatic influences which may operate on the several routes.

Upon your application, the Finance Minister will make all necessary arrangements with regard to the supply of funds.

I shall feel obliged by your transmitting information from time to time touching the the progress of your survey.

I have the honor to be, Sir,
Your obedient servant,
(Signed,) JOHN SIMPSON

S. FLEMING, Esquire,
Quebec.

REPORT.

The Exploratory Survey of 1864, conducted by me agreeably to the foregoing instructions and correspondence, has been brought to a close, and it now becomes my duty to report the result.

The main object of the Survey was to enable the Government to judge of the comparative merits of the various routes which have been proposed, as well as any other routes which seemed worthy of attention and feasible for a Railway to connect the Provinces of Nova Scotia and New Brunswick with Canada.

A Railway is already in operation from Halifax, the capital of Nova Scotia, northerly to Truro, in length 61 miles; and the Canadian Railway system extends to River du Loup. The portion of the contemplated International Railway remaining to be constructed lies therefore between Truro and River du Loup.

The distance between Truro and River du Loup by an air line is about 360 miles, and the width of the country within which various routs for the Railway have been proposed, averages not less than 100 miles, much of it moreover is covered with a dense unbroken forest; it is evident therefore that in a field so extensive and so difficult to penetrate, that full justice to the important enquiry could scarcely be expected to be done in one short season.

It was, however, the urgent desire of the Government that they should be placed in possession of such information as might result from the survey at the very earlist period; I therefore took measures to prosecute the work energetically and to carry out as much of the instructions as it was possible to do within the very limited time which has elapsed since the exploration commenced.

The winter of 1863-64 had commenced before I was fully authorized to proceed with this important service.

I began by making a reconnoissance of the country within the limits of the survey, at least so far as this could be done by travelling rapidly over the roads that were opened, and on the rivers that were passable at that season of the year. At the same time, I instituted barometrical explorations across the Tobique highlands from Boiestown northerly; as well as on the height of land between the Restigouche and the St. Lawrence.

A large quantity of provisions were also forwarded on the snow and stored at a convenient point in the interior of the country, for the future use of surveying parties.

These necessary preliminary services were completed by the close of winter; immediately thereon four efficient surveying parties were organized, ready to take the field on the snow leaving the ground, or so soon thereafter as circumstances would admit, and to continue at work simultaneously, during the season to the completion of the survey.

THE ENGINEERING STAFF.

To assist me in this survey I selected gentlemen who were previously well known and who have since proved to be eminently qualified for the several duties assigned to them.

An experienced engineer was placed in immediate charge of each surveying party, whose duty it was to carry out my wishes and direct the assistants and men under him.

Each surveying party besides the engineer in charge, consisted of a sufficient number of assistants to carry on the levelling, surveying and barometrical observations together with a full complement of axemen and packmen.

Besides the men immediately connected with the surveying parties, Indians and others, were engaged to aid in exploring and also in forwarding supplies to the interior of the wooded districts, during the prosecution of the survey.

The first party left Quebec in charge of Walter Lawson, Esq , C E., on May 25th, and proceeded immediately to the highlands where the Rimouski, the Kedgwick (a tributary of the Restigouche), the Green River (a tributary of the St. John), the Toledi and other rivers take their rise.

The second party left Quebec in charge of Thos. S. Rubidge, Esquire, C. E., on the 28th of May, and proceeded by the Temiscouata road to Little Falls on the St. John River, Thence by the Grand River and Wagan portage to the River Restigouche. This party commenced operations by tracing up the Gounamitz River from its confluence with the Restigouche.

The third party left Quebec with myself, on the 31st of May, by the provincial Steamer "Lady-Head" for Dalhousie. Samuel Hazlewood, Esquire, C. E., was placed in charge of this party, and he began the season's operations by making an exact survey of the River Metapedia from the Restigouche upwards.

David Stark, Esquire, C. E , took charge of the fourth party; he left Quebec on the 14th of June, by the " Lady Head " for Nova Scotia. He commenced the survey in that Province by tracing a line through a Gap in the Cobequid range, previously discovered to the north of Parsboro', and thence he afterwards continued the survey in the direction of Truro.

Soon after these several parties left Quebec, they were actively engaged in the field, and throughout the season nearly one hundred persons in all were employed in connection with the survey. This force, with little change and no intermission continued at work in the woods until the close of field operations late in November.

Various kinds of flies were more than usually troublesome during the first half of the season. The parties engaged in the northern section of the country suffered very much.

Since the close of operations in the field, the engineering staff have been actively engaged reducing the survey to paper.

MAIN DIVISIONS OF THE SURVEY.

An air line drawn between Truro, the nearest point of connection with the Nova Scotia Railway, leading to Halifax and River du Loup, the eastern extremity of the existing Canadian Railway system, is in length about 360 miles; it crosses Cumberland Basin and the Petitcodiac Inlet, both navigable extensions of the Bay of Fundy. These waters cannot be crossed on an air line, and therefore to avoid them it becomes necessary to keep some distance easterly, as far at the very least as a point known as " The Bend of the Petitcodiac," from this point an air line drawn to Truro will clear Cumberland Basin.

Between the tidal waters of the Bay of Fundy at the Bend of the Petitcodiac, and the waters of the Gulf of St. Lawrence at Shediac Harbor, the distance is only 13 miles, and within the limits of this narrow isthmus any railway from the mainland to the Peninsula of Nova Scotia must necessarily pass. The consideration of the whole question of route very naturally, therefore, is divided into two main divisions by the conformation of the country here alluded to. A Railway is constructed across the Isthmus from Shediac to Moncton, a small town at " The Bend," thence westward to the City of St. John, New Brunswick ; and as this Railway in part forms a section of some of the contemplated Intercolonial Railway routes, it seems convenient to make it the separating line between the two divisions of the survey, in which, at present, it is proposed to consider the subject. South of the New Brunswick Railway will therefore, in the following, be called the " Nova Scotia Division," and north of this Railway the " New Brunswick and Canada Division " of the survey.

THE NOVA SCOTIA DIVISION OF THE SURVEY.

The chief obstacle to be overcome on this division of the survey is a range of highlands known as the Cobequid Hills, lying immediately to the north of Truro. This conspicuous range seems to divide the Bay of Fundy into two great forks, the most northerly one some fifty miles in length, and terminating in the Cumberland Basin, at the head of which is the town of Amherst, the more southerly fork not less than eighty miles in length, from Cape Chignecto to the head of Minas Basin at Truro.

The Cobequid Hills range in altitude from 800 to 1,000 feet above the sea; they extend almost due east and west of Truro, to a total length of about one hundred miles, and with a breadth averaging perhaps about ten or twelve miles. Moncton is nearly north-west from Truro, and, therefore, the general direction of the Railway route crosses the Cobequid range obliquely.

North of the Cobequid Hills the surface of the country is comparatively flat; at one or two points it is irregular and broken, but no difficulties of an unusual character occur.

At different times four lines have been surveyed from Truro towards New Brunswick; beginning with the most easterly, they may be briefly described in the following order :—

Line No. 1.—From Truro this line runs easterly along the valley of the Salmon River, following the route of the Railway now under construction to Pictou, to a place known as Wall's Mill, some ten miles out of Truro; thence it turns northerly and crosses the Cobequid range in the neighborhood of Earltown, at an elevation above the sea of 506 feet; descending to the general level, it then runs to the west of Tatmagouche, Wallace and Pugwash, generally parallel to the Gulf coast to the boundary of New Brunswick at Bay Verte; thence, prolonged northerly, this line was intended to intersect the Railway from St. John to Shediac near the latter place. This line was surveyed about the year 1853, by Mr. James Beatty for an English contracting firm. I believe it was found generally favorable with gradients, except on the northern slope, not exceeding 53 feet per mile, and minimum curves of half a mile radius.

Line No. 2.—This line runs from Truro in a north-westerly direction up the southern slope of the Cobequid range until it reaches Folly River, following which the summit is attained at Folly Lake, at an elevation of 600 feet above high tide water. Folly Lake is situated in a pass through the high lands, within which Folly and Wallace Rivers take their rise; the former flowing southerly, the latter northerly.

The descent of both streams is very rapid, involving heavy work and heavy gradients, the latter ranging from 60 feet per mile for about six miles ascending northerly, to 66 or 70 feet per mile, descending on the opposite side. Some lesser difficulties occur to the north of the main range, but after the River Philip is crossed the country undulates easily, and the line will then be direct with favorable gradients.

This line was surveyed under the directions of the late Major Robinson, in 1847, and described in the Report of Captain Henderson.

Lines Nos. 1 and 2 are common north of Bay Verte.

Line No. 3.—This line follows the same general direction as line No. 2, until the Folly River is reached, but instead of turning to the north and crossing through the Folly Pass, it continues ascending the southern slope of the high ground to a stream known as Great Village River. After crossing a branch of this stream by an expensive viaduct the line strikes the main valley near the Acadian Mines, and continues along the eastern bank on an ascending gradient to the summit at Sutherlands Lake, 24 miles out of Truro, and 700 feet above the sea. The heaviest gradient between Truro and the summit is about 62 feet per mile for 1½ miles, and extends from the Acadian Mines upwards.

The descent on the northern slope is comparatively easy, the gradients not exceeding 53 feet per mile. After crossing the Cobequid range, the line continues in a direction north-westerly to Amherst, Sackville, Dorchester, and thence to a point on the St. John and Shediac Railway, about six miles easterly from Moncton. This line has not been instrumentally surveyed for a distance of over 30 miles, between Sackville and the River Philip, 41 miles from Truro, but the country is favorable and no serious difficulty is apprehended. Between Sackville and Moncton, the only obstacle of any moment is a high ridge near Dorchester. The profile on the line surveyed shows ascending and descending gradients at this point of about 80 feet per mile, but I am induced to think that farther surveys may prove that these heavy gradients need not be adopted.

The portion of this line extending 41 miles out of Truro was surveyed during the past year by Alexander Beattie, Esquire, C. E., for the proprietors of the Acadian Mines, the section lying between the Provincial Boundary line near Amherst, and Moncton, about 33 miles in length, was surveyed last year by J. E. Boyd, Esquire, C. E., under instructions from the Government of New Brunswick.

The following is an abstract of the aggregate length of grades shown on the profiles:

From Moncton to Tantramar River.

		Ascending Southerly.	Ascending Northerly.
Grades under 20 feet to the mile		2.9 miles.	3.1 miles.
" 20 to 30 feet to the mile		1.1 "	1.6 "
" 30 to 40 " "		1.5 "	0.9 "
" 40 to 50 " "		0.7 "	2.5 "
" 52-8 " "		0.7 "	2.2 "
" 79 " "		2.3 "	0.7 "

Level.. 10.1 miles.
Total length of Section .... 30.3 "

From Truro to River Philip.

	Ascending Southerly.	Ascending Northerly.
Grades under 20 feet to the mile	1.4 Miles.	0.4 miles.
" 20 to 80 " "	1.5 "	1.4 "
" 30 to 40 " "	2.3 "	0.0 "
" 40 to 50 " "	0.0 "	1.5 "
" 52-8 " "	11.8 "	4.6 "
" 59 " "	0.0 "	4.8 "
" 62 " "	0.0 "	4.3 "

Level.. 7.1 miles.
Total length of section .... 44.1 miles.

Line No. 4—Nearly due South of Amherst a break or opening in the Cobequid range occurs, and presents a very favorable opportunity for crossing from the head of the northerly fork of the Bay of Fundy to the Basin of Minas at the head of which Truro is situated. In this opening a branch of Macan River, which flows into Cumberland Basin, near Amherst, and also Partridge River, which flows into Minas Basin near Parsboro', take their rise. The summit between these streams is less than a hundred feet above high tide and suggestive of every easy gradients. In every other respect the ground for 30 or 40 miles southerly from Amherst is extremely favorable for a Railway line. The same may be said of the country for a like distance on the southerly end of this line, viz.: from Truro to a place called Economy, along the coast of the Basin of Minas. From Economy to Parsboro' the survey did not prove so satisfactory. Two spurs of the Cobequid range had to be surmounted; the one at a level of 350 feet and the other at 230 feet above high tide water. Several deep ravines had also to be crossed, involving heavy work on this section; and the maximum gradients found necessary between Parsboro' and Economy, ascending and descending, are 60 feet per mile.

The approximate profile prepared from the Exploratory Survey made under my direction during the past season, from Jeffers Lake, a few miles north of Parsboro', to Truro, has the gradients laid down thereon, of which the following is an abstract:

			Ascending Southerly.	Ascending Northerly.
Total length of Grades under 20 feet to the mile			8.5 miles.	5.1 miles.
" " 20 to 30 "			6.5 "	4.2 "
" " 30 to 40 "			2.2 "	4.7 "
" " 40 to 50 "			0.0 "	1.7 "
" " 52.8 "			2.2 "	5.0 "
" " 60 "			5.1 "	1.9 "

" Level .... 12.9 Miles.
Total Length of section................................. 60.0 "

From Jeffers Lake northerly to Amherst and the New Brunswick boundary, the country is so simple in its features that a survey was not deemed necessary. From Amherst, northerly, lines Nos. 3 and 4 are common. The lengths of these four lines from Truro to a common point east of Moncton, according to the best information in my possession, may be given as follows:

Line No. 1.

From Truro along Pictou Railway under construction to Walls Mill....	10 miles.
From Walls Mill to intersection with New Brunswick Railway near Shediac	106 "
From intersection, near Shediac, along New Brunswick Railway to point east of Moncton	7 "
Total............	123 miles.

Of which 17 miles are already constructed or in progress.

Line No. 2.

From Truro to intersection with New Brunswick Railway, near Shediac.	103 miles.
From intersection near Shediac along New Brunswick Railway to point east of Moncton..............	7 "
Total............	110 miles.

Line No. 3.

From Truro, by Acadian Mines and Amherst, to point east of Moncton.   106 miles.

Line No. 4.

From Truro, by Parsboro' and Amherst, to point east of Moncton.......   125 miles.

A fifth line may be had by connecting line No. 1, after crossing River Philip, with lines No. 3 and 4 in the neighbourhood of Amherst, and a sixth line may be had by combining lines Nos. 2 and 3, by a short connection running from the former near Tullocks Creek, to the latter near Salt Springs.

The total length of No 5 would be about............... 124 miles.
         Do   of No. 6   do   do... .... 111 "

And the several lines, so far as distance is concerned, would stand thus:

No. 1 —123 miles, Truro to point east of Moncton, by Shediac.
No. 2.—110 miles, ... do .... by Shediac.
No 3.—106 miles, ... do ....
No. 4.—125 miles, ... do ....
No. 5.—124 miles, ... do ....
No. 6.—111 miles, ... do ....

The greatest length of level or easy gradients will be found on line No. 4, whilst on lines Nos. 1 and 5 will be found the lowest maximum gradients. In this respect, line No. 3 next appears most favorable, but in making a comparison between these different routes, it becomes necessary to exclude the heavy ascending and descending gradients common to lines Nos. 3, 4, 5 and 6, near Dorchester

The obstacles in this quarter can certainly be overcome with easier grades either by an increase of cost or of distance, for which ample allowance will be made in the estimate. It appears that lines Nos. 2 and 6 crossing the Cobequid ridge by Folly pass have the least favorable gradients.

Lines Nos. 1 and 2 would best serve the local traffic at present centering in the villages of Tatmagouche, Wallace, Pugwash, and Bay-Verte on the Gulf coast.

Line No. 3 would accommodate Amherst, Dorchester and Sackville. And *Line No. 4*, in addition to serving these points, would also accommodate Parsboro' and the several villages along the north shore of the Basin of Minas.

Line No. 5 would equally with No 1 serve Tatmagouche, Wallace and Pugwash, whilst at the same time it would pass through Amherst, Dorchester and Sackville.

Line No. 6, whilst passing through Amherst, Dorchester and Sackville, would, to the same extent as line No. 2, accommodate the population on the Gulf shore around Tatmagouche, Wallace and Pugwash.

The country south of Amherst on the Macan River and some of its tributaries, abounds in coal in thick beds and of excellent quality. This valuable coal field would be opened up by lines Nos. 3, 4 and 6.

The Cobequid range is rich in iron ore of the best description; it is now manufactured on the southern flank of the range, at the establishment of the Acadian Iron Company. Annually, considerable quantities of iron are exported to England, and there converted into steel, for which, from its quality, it is admirably adapted. It is considered that iron manufactures of all kinds would be established and greatly multiplied in this section, were proper facilities created for bringing the coal and ore together. Line No. 3 accomplishes this end, and so also does Line No. 6; although the latter does not in the same degree accommodate the existing establishment of the Acadian Mining Company, now in operation on Great Village River.

In review of the above, it would seem that, apart from the question of distance and gradients, a central route, whilst opening up the mineral districts both of coal and iron, would at the same time serve generally the population of the country as well as any other line specially located with that object solely in view, and without regard to the development of the rich mineral resources of this district.

Although the surveys which have been made show that the central routes referred to are the shortest, they have not the advantage when gradients are considered, still I am convinced that further surveys would result in modifying and greatly improving one or other of these lines, or in finding, in part at least, a new line which, whilst securing all the advantages claimed for either of the central lines, would have the additional recommendation of possessing more favorable gradients and curves throughout, from Truro to Moncton. It would not be wise to calculate that an improved central line can be had, without to some extent affecting the cost and distance. I shall, therefore, in the estimate consider the distance from Truro to the point intersected with the New Brunswick Railway, east of Moncton, as 109 miles, nearly a mean between the length of line No. 6 and No. 3; thus making ample allowance for the improvement of the gradients at Dorchester, as well as those on the ascent to the Cobequid summit, should the general route of line No. 3 be finally adopted.

Between Moncton and Truro, with the exception of the mineral districts which are for the most part in a state of wilderness, much of the country is settled, and in some sections cultivated farms of the richest description can be seen.

ESTIMATE OF QUANTITIES.

I shall now proceed to give the quantities of the principal kinds of work required to complete the bridging and grading on the sections surveyed last summer. These quantities are the data on which I shall base the estimate of cost when I come to that part of the subject; they are calculated from the profiles of the lines which have been made from the information derived from the survey; but as the profiles are, in some cases at least, only approximate, great accuracy cannot be expected. Tables have been prepared, showing the quantities of work on each separate mile, of which the following is a summary:—

From the point of connection with existing Railway, east of Moncton, to Tantramar River, near Sackville, length of line surveyed, 30 miles.

1. Common Excavation...1,083,854 c. yards.
2. Assumed proportion of Rock Excavation.....................   114,146   "

                Total Excavation......................1,198,000   "
3. Culvert Masonry...   10,771   "
4. Bridge   do.   2,132   "
5. Weight of Wrought-Iron Bridges.................................   435 tons.

From Truro to East Branch of River Philip near Rufus Black's, by way of the Acadian Mines. Length of this section as surveyed 41$\frac{13}{100}$ miles.

1. Common Excavation..1,945,000 cubic yards.
2. Assumed proportion of Rock Excavation................   586,000   "
                                                                                                        ———2,531,000 c. yds.
3. Culvert Masonry...   27,023   "
4. Bridge Masonry..   13,272   "
5. Weight of Wrought-Iron Bridges..........................   ⁻ 876 tons.

Between Tantramar River, where the first section above referred to ends, and Rufus Black's, on the River Philip, where the second section begins, an instrumental survey has not been made, and, in consequence, there is no certain data from which the exact quantities of work can be computed. It is believed, however, that the following rough estimate, from a hurried examination of this intermediate section, will, when added to the above quantities, give a full estimate of the work on the whole line between Moncton and Truro.

1. Common Excavation..894,000 cubic yards.
2. Assumed proportion of Rock Excavation....................   7,000   "
                                                          ———901,000 c. yards.
3. Culvert Masonry...   12,000   "
4. Bridge   do.   7,650   "
5. Wrought-Iron in Bridges..   436 tons.

Adding the quantities above given together, we shall then have the total quantities of the chief kinds of work required to complete the bridging and grading of the whole line within the Nova Scotia Division of the survey; that is to say, from Moncton to Truro, as follows:—

1. Common Excavation..3,922,854 cubic yards.
2. Assumed proportion of Rock Excavation....................   707,146   "

                Total Excavation..4,630,000   "
3. Culvert Masonry...   49,791   "
4. Bridge   do.   23,054   "
5. Bridge Iron ....   1,747 tons

The quantities on the line by way of Parsboro' (No. 4) have been computed in a manner similar to that above described with the following results:—
1. Common Excavation..4,765,954 cubic yards.
2. Assumed proportion of Rock Excavation....................   388,146   "

              Total Excavation..5,154,100   "
3. Culvert Masonry...   44,634   "
4. Bridge   do.   20,702   "
5. Weight of Iron in Bridges.....................................   1,877 tons.

In calculating the quantities of earthwork, in every case the cuttings have been estimated 30 feet wide at formation level, side cuttings 24 feet, and embankments 18 feet wide; the various structures are intended to be of a substantial and permanent character, they are estimated to be either stone Culverts, or Bridges made of wrought iron on stone abutments and piers, and it is believed that the quantities herein given are ample.

The probable cost of this division of the work will be considered when that of the whole line is taken up.

NEW BRUNSWICK AND CANADA DIVISION OF THE SURVEY.

Two Railways are already constructed and in operation within the limits of the Province of New Brunswick; one, designated the New Brunswick and Canada Railway, commences at the Town of St. Andrews on Passamaquoddy Bay, at the extremely south-westerly angle of the Province; it extends in a northerly direction, parallel to and not far from the boundary of the State of Maine, a distance of nearly ninety miles, to a point known as Richmond Station, some four or five miles to the west of the Town of Woodstock. The other line in operation is designated "The European and North American Railway." It begins at the city of St. John on the north shore of the Bay of Fundy, and extends a distance of about 105 miles, in a north-easterly direction, to Shediac, on the Gulf of St. Lawrence. In considering the subject of Intercolonial communication, two points on this line of Railway are of great importance; one, the City of St. John, although not the political capital, the commercial centre of New Brunswick, and the other, Moncton, which commands every possible overland route, not only from Canada and New Brunswick, but from the United States to Nova Scotia, and its capital, Halifax.

St. John, although the great commercial centre of New Brunswick, is not, however, the only place of importance. There are towns, such as Frederickton, the seat of Government, Woodstock and other places on the western side of the Province; and Chatham, Bathurst, Dalhousie and Campbelltown on the Gulf coast. These all possess a certain amount of local traffic, the accommodation of which it is desirable to keep in view. It unfortunately happens, however, that a line constructed in River du Loup by the coast to Moncton, whilst best serving Halifax and the population on the east of New Brunswick, would do so at the expense of St. John and other places in the west.

It will be seen, too, that a direct line from St. John would serve that city and the towns and settlements in the west, whilst the points referred to on the Gulf coast would necessarily be neglected.

This is here alluded to in order to show that the selection of a Railway route through New Brunswick, is involved in local sectional difficulties at the very outset. The settlement of the Province has naturally enough followed its navigable waters; on the south by the Bay of Fundy and its inlets; on the east by the coast and bays of the Gulf of St. Lawrence; and on the west by the River St. John, which extends, and to some extent is navigable, almost to the extreme north-westerly angle of the Province. In consequence, New Brunswick may be said to be peopled as yet only round its outskirts. There is a vast area in the interior unoccupied, not because the soil is so much more uncultivable than elsewhere, but because it had hitherto been, and is still, inaccessible.*

Although I have chiefly to deal with the engineering features of the subject, these considerations cannot be overlooked in taking up the whole matter covered by my instructions, as in view of traffic for the contemplated Railway, the question of route is very naturally and very properly influenced by the present and prospective business of the country traversed.

An air line drawn from the City of St. John to River du Loup, in about 250 miles in length, but such a line falls within the State of Maine, as much as 25 miles. The shortest line that can be drawn on British territory, is some five miles longer; it extends directly from St. John to the north-easterly angle of Maine near the Grand Falls, thence along the boundary some thirty miles, then straight across the country by Little Falls to River du Loup.

An air line drawn from Moncton to River du Loup, passes entirely within British soil; although near Little Falls, it comes within two or three miles of the American boundary— this line is 260 miles in length.

Practically then, the relative position of these three points, viz.: River du Loup, Moncton and St. John, may be viewed as forming the angles of an isosceles triangle, the base of

---

* "A Parallelogram bounded on the south-east by a line drawn from Frederickton to Chatham, on the north-east by a line drawn from Chatham to Metis, on the south-west by a line drawn from Frederickton to River du Loup, on the north-west by the settlements along the River St. Lawrence; about 90 miles in width, by about 200 miles in length, and embracing nearly 18,000 square miles, is both unsettled and roadless."

which is the Railway in operation from St. John to Moncton, 90 miles, and the sides from 255 to 260 miles in length.

The construction of a Railway on either of these direct lines is quite impracticable; there are many engineering difficulties on each, which render it necessary to depart materially from the straight course; and if practicable, for military reasons the building of an Intercolonial Railway on either of these lines, touching, as they do, the American frontier, is pronounced by military authorities objectionable.

In seeking to avoid the great military objection to any line in close proximity to the American boundary, we unfortunately increase the engineering difficulties; as, in looking for a line sufficiently distant from the frontier, unless we at once go to the other side of the Province, and thus considerably increasing the length, we are driven into a section of the country characterised by great irregularities of surface and difficult to penetrate.

In dealing with the whole subject we cannot, however, overlook military considerations, and although it is difficult to learn exactly what minimum distance from the frontier would satisfy the military authorities, reference to this question is unavoidable.

I could not presume to express an opinion on the best military position for the Railway, or even enter into the question of route in a purely military aspect at all; but in the absence of any specific instructions or suggestions on this point, I found it necessary to look for some rule by which to be guided at the beginning and during the progress of the survey. For a number of miles west of River du Loup, the Grand Trunk Railway passes the northwestern boundary of the State of Maine at a distance of scarcely 30 miles; this, at all events in a military aspect, is a precedent, and may suffice to establish the minimum distance allowable between the contemplated line of Railway and the north-eastern angle of the same State. I have accordingly laid off this distance on the accompanying general map of the country, from the frontier to points on the River Trois Pistoles, Green River, the Restigouche and Tobique. Lines connecting these points and prolonged direct to St. John in the one hand and to Moncton on the other, may, simply to distinguish them from other lines, be termed "Military air lines."

These "Military air lines" (so called) are intended not to approach the American frontier at any point nearer than the Grand Trunk Railway does in its course between River du Loup and Quebec.

Such lines connecting River du Loup with St. John measure about 273 miles, and from River du Loup to Moncton, about 265 miles.

While having due regard to routes which, for their commercial or engineering reasons simply, might approach or touch either the American frontier on one side of New Brunswick, or the Gulf coast on the other, I ventured to assume that the military authorities would offer no decided objection to the construction of the contemplated Railway on or near the lines last referred to.

I had in view, therefore, from the beginning of the survey, the discovery of at least one practicable route for the Railway, which, without increasing the distance unnecessarily, would conform, as near as possible, with the guiding rule above alluded to.

A section of the country on either of these Military air lines, whilst showing that the construction of a Railway precisely thereon is entirely beyond the limits of practicability, will, at the same time, indicate and illustrate the bold physical features which characterise a very large portion of the territory embraced by the survey.

Beginning at River du Loup and following the line laid down at the prescribed distance from the Maine boundary to the City of St. John; we find that in passing over the mountainous ridge which separates the St. Lawrence from the Restigouche, not only is a maximum elevation of nearly 2,000 feet above the sea reached, but the surface passed over is of a very broken character; minor ridges nearly all crossing the line in a right angled direction, are constantly met with; these attain elevations ranging from probably 1,000 feet to nearly double that height above the sea, and are separated by low lying water channels, of which may be mentioned, Lake Temiscouata, River Toledi, Squatook Lakes, besides the branches of Green River. Several of these waters will not exceed 500 feet above sea level.

The distance from River du Loup by the air line at its crossing the Restigouhe River is nearly one hundred miles, and the latter river at the crossing is about 450 feet above the sea. The great ridge continues easterly between the St. Lawrence on the north, and the Restigouche and Bay Chaleurs on the south, until it terminates in the Gaspé Peninsula. It must be crossed at some point by any line of Railway communication, intended to connect the Maritime Provinces with the Canadas, but the section now being described crosses it in perhaps one of the least favorable directions.

Continuing from the Restigouche southerly to Tobique, a distance of about 35 miles, the line crosses a heavy irregular swell running easterly and westerly, and attaining a summit height varying from 1,000 to 1,200 feet above the sea. The line crosses the Tobique at about 500 feet above the same level. From the River Tobique continuing southerly it has a third main ridge to cross; this ridge is known as the Tobique Highlands, it extends easterly from the River St. John to a rugged district in the interior of New Brunswick, where the Tobique, the Upsalquitch, the Nepisiguit, and some tributaries of the Miramichi take their rise. On the air line from St. John, this ridge separates the Tobique from the main Miramichi, and is, in a direct line, about 45 miles in width; the height of land passed over will probably not be less than 1,500 or 1,700 feet. The height of the River Miramichi at the crossing is probably a hundred feet greater than at the Tobique crossing.

South of the Mirimachi on the same line continued, the ground rises again to a considerable elevation and is intersected by deep river valleys. The line passes to the east of Fredrickton some eight miles and crosses the River St. John about twelve miles below that city. Continuing onwards it crosses the River a second time, as well as a long, wide and deep extension of the St. John River called Kennebecdasis Bay, besides a good deal of broken ground immediately north of the city of St. John.

The (so called) Military air line, from River du Loup to Monctou, passes over ground north of the Miramichi, not dissimilar to that of the St. John air line above described. The country between the Miramichi and Moncton is much simpler in its character, and on this section no insurmountable difficulties exist.

Aware of the importance of a favorable Railway route in the general direction of the military air line above alluded to, I determined to exert every effort to discover one; although it must be confessed the above sketch of the leading features of the country, and the following extracts from the report and correspondence of Major Robinson, dated 1848 and 1849, made it appear extremely doubtful that a practicable line could be had.

" The fourth obstacle is the broad and extensive range of highlands which occupies nearly the whole space in the centre of New Brunswick, from the Miramichi River north to the Restigouche. Some of these mountains rise to an attitude exceeding 2,000 feet.

" The Tobique River runs through them, forming a deep valley or trough which must be crossed by the direct line, and increases greatly the difficulty of passing by them.

" The lowest point of the ridge overlooking the Tobique River, at which any line of railway must pass, is 1216 feet above the sea. Then follows a descent to the river of 796 feet in 18 miles, and the summit level on the opposite ridge or crest between the Tobique and Restigouche waters 920 feet above the sea, or a rise of 500 feet above the point of crossing at the Tobique water. These great summit levels which must be surmounted, form a serious objection to this route."

*   *   *   *   *

" The fifth and last obstacle to be overcome, and which cannot be avoided by any of the routes, is the mountain range running along the whole course of the River St. Lawrence in a very irregular line, but at an average distance from it of about twenty miles. It occupies with its spurs and branches a large portion of the space between the St. Lawrence and the Restigouche Rivers. The rocks and strata composing the range are of the same character and kind as the Tobique range. The tops of the mountain are as elevated in the one range as in the other.

" The exploring parties failed in finding a line through this range to join on to the direct line through New Brunswick, but succeeded in carrying on the Eastern or Bay Chaleurs route, owing to the fortunate intervention of the valley of the Metapediac River.

" The line which was tried and failed was across from the Trois Pistoles River, by the heads of Green River and down the Pseudy or some of the streams in that part running into the Restigouche River."

*   *   *   *   *

4

"From Boiestown the general course was followed, and levelled as far as the Tobique River, but the country was so unfavorable that new courses had to be constantly sought out.

"A new line altogether was tried from the Tobique as far as the Wagan portage.

"The results deduced from the observations and sections proved this line to be quite impracticable for a Railway.

"Whilst the line was being tried, other parties explored from Newcastle on the Miramichi River, over to Crystal Brook on the Nipisiguit, the vallies of the Upsalquitch and its tributaries and as far as the Restigouche River.

"The country at the upper waters of the Nipisiguit, and the whole of the Upsalquitch valleys, were found to be rough, broken and totally impracticable.

"The result of this season's labours went to show that the best, if not the *only* route that would be likely to be practicable, would be by the North-west Miramichi to Bathurst, and then along the Bay Chaleurs."

* * * * *

"A large party was engaged in trying to find a line from Trois Pistoles River on the St. Lawrence through the Highlands to the Restigouche River, for the purpose of connecting on to the New Brunswick party. The winter overtook them whilst still embarrassed in the Highlands at the head waters of the Green river.

"The dotted lines on the General Plan will show their attempts.

"A line was tried up the valley of the Abersquash, but it ended in a *cul-de-sac*; there was no way out of it.

"A second line was carried from Trois Pistoles over to Lac-des-Iles, Eagle Lake; and by the middle branch of the Tuladi River, the north west branch and head waters of the Green River were gained.

"But this point was not reached except by a narrow valley or ravine of four miles in length.

"A Theodolite section was made of it, and it was found to involve a grade of at least one in forty-nine, and to attain that, heavy cuttings at one part and embankments at another would be necessary.

"There is no occasion at present to enter upon the discussion of whether this should condemn a whole line; for having attained the Forks at the head of the main Green River, no way was found out of it, and this explored line, like the first mentioned, must be considered to have ended in a *cul-de-sac* also."

* * * * *

"Large parties were thus employed at great expense for two seasons on this central and direct line through New Brunswick.

"Judging from the results of our labours, from those of others, and the natural difficulties of the country as described, I do not think any further explorations would be attended with any marked difference of success."

The exploration undertaken on snow shoes, early last year from Boiestown on the Miramichi northerly to the river Tobique (together with information from other sources) resulted so far satisfactory, that no obstacles of an insuperable nature were apprehended in that quarter.

The exploration similarly undertaken between the St. Lawrence and the Restigouche during the winter 1863-64, although it added to the information previously gathered, proved unsuccessful in the main object in view; and in consequence, the probability of finding a practicable passage for the Railway, between these waters, was rather diminished than increased by the additional knowledge of the country thus obtained.

Hence it appeared of the utmost importance, to have this section carefully explored, before commencing the Railway survey on any other portion of a direct central route; so soon as this vital point became thoroughly understood, it would then be easy to decide whether to proceed with or abandon the survey through the interior.

Vigorous measures were required to settle the question of practicability through this district with as little delay as possible. I, therefore, concentrated the efforts of two thoroughly efficient and well appointed surveying parties to the solution of the difficulty.

One party entered on the exploration from the Restigouche, following up the valley of the Gounamitz, and aiming at the discovery of a passage into the valley of Green River, near its south-easterly source.

Another party entered from Rimouski, with the view of finding a suitable passage from the valley of Rimouski River, by its south-easterly branches to the valley of the Kedwick, and thence, should the first mentioned party fail, to the River Restigouche.

Both attempts proved successful.

Having thus a choice of routes across the height of land forming the northerly water shed of the Great Restigouche Basin, and being unable from the shortness of the season, and more particularly from the very limited appropriation at my command, to follow up both, it became necessary to make a selection; I therefore decided reluctantly to abandon the exploration by the Rimouski and Kedwick, and determined to continue the survey by the Gounamitz and Green River; the latter route appearing the most direct, and at the same time sufficiently remote from the frontier. On arriving at this decision, both parties were placed on the Gounamitz route.

Whilst these explorations were in progress, two other equally efficient surveying parties were engaged, the one in Nova Scotia, between Truro and Moncton, the other in making a re-survey of that portion of the line through the Matapedia valley, considered the most difficult and expensive of the route recommended by Major Robinson. The character and results of the latter examination will hereafter be referred to.

So soon as the party in Nova Scotia had completed all that I felt justified in doing in that Province, I immediately transfered it to New Brunswick, and there engaged it in the continuation of the line which commenced in the valley of the Gounamitz.

Anxious to have a continuous instrumental survey, from the St. Lawrence to the line of railway running from St. John to Moncton, before the season closed and the appropriation became exhausted, I transferred the Matepedia party, early in October, to the south of New Brunswick to aid in this work. From the beginning of October to the close of the field operations, the four parties were simultaneously engaged on the same route.

By the beginning of December, a continuous line of levels and other measurements were made from Trois Pistoles to Apohaqui Station, about midway on the railway running from the city of St. John to Moncton. And thus, although the object of the survey was mainly to assertain beyond a doubt, that there was nothing impraticable in the way; yet the additional information obtained, by the completion of the instrumental measurements on this particular line, is doubtless of very considerable importance, as it gives pretty satisfactory data on which to base an approximate estimate of the probable cost of the line surveyed; as well as collateral data of some value, in estimating the cost of other possible lines, through analagous sections of the same country, but which as yet have not been similarly examined.

THE SURVEYED GENERAL LINE. *

I shall now proceed to give an outline of the engineering and other features of the Central Route above referred to, beginning at the point of connexion with the Grand Trunk Railway near River du Loup, and terminating at Apohaqui Station, on the New Brunswick Railway.

I found that an exploratory survey had been made some six years ago, in connection with the works of the Grand Trunk Railway from River du Loup easterly to River Trois Pistoles, a distance of 24 miles. This survey was of a satisfactory nature, and it was therefore deemed unnecessary to go over the same ground a second time.

RIVER DU LOUP TO RIVER TROIS PISTOLES.

On this section three rivers of importance are crossed, viz.: River du Loup, River Verte, and River Trois Pistoles. The last will require a bridge of great magnitude, as the river flows in a rocky gorge about 150 feet deep and of considerable width even at the most favorable point. It is proposed to cross this river and ravine on a viaduct of thirteen spans, one of which is intended to be 100 feet in the clear, and the remaining

---

* (New Brunswick and Canada Division of the survey.)

twelve with 60 feet openings. The bridges over the Rivers du Loup and Verte will each have three 70-feet spans. The former will be about 22 feet above the water, and the latter 30 feet.

The following summary of the grades given on the profile will show that they are on this section extremely favorable, very few being over 40 feet to the mile; the highest ascending south is about half a mile in length at 52.8 feet to the mile, and the maximum ascending north is 53.5 feet per mile.

CHARACTER OF GRADES.	TOTAL LENGTH OF GRADES IN MILES.	
	Ascending South.	Ascending North.
Under 20 feet per mile	8.4	4.6
20 to 30 do	0.3	0.0
30 to 40 do	2.5	0.0
40 to 50 do	0.9	1.0
51.9 to 52.8 do	2.4	0.0
53.5 do	0.0	1.4

Level.. 3.0 miles.
Total length of Section................................... 24.5 miles.

The quantities of the chief kinds of work, which the profile shows as necessary to complete the bridging and grading, in an efficient manner on this section are as follows:

1st. Common excavation........................ 484,289 cubic yards.
2nd. Assumed proportion of rock excavation 39,635 do do
   Total excavation..................... 523,924 do do

3rd. Culvert masonry............................. 4,016 cubic yards.
4th. Bridge masonry.............................. 6,961 do
5th. Weight of Bridge Iron.................... 414 Tons.

RIVER TROIS PISTOLES TO GREEN RIVER FORKS.

Beginning above the confluence of the River Abawisquash with the Trois Pistoles, at an elevation of 497 feet above tide water, the line follows the valley of the Abawisquash, with grades not exceeding 50 feet per mile for a distance of eleven and a half miles; here it passes over a summit only 690 feet above the sea, into the Basin of Island Lake; descending gradually from the water shed between the Abawisquash and Island Lake, for a distance of about eleven miles with remarkably easy grades, seldom over 15 feet per mile, it reaches the head of Eagle Lake, 532 feet above the sea. The line surveyed now turns in an easterly direction and ascends to the Wagan Lake, 50 feet above and four miles distant from Eagle Lake. It then curves on a perfect level to the valley of the Turadi, a tributary of the Rimouski, and following the valley of the former with nearly level, or grades under 20 feet to the mile, it reaches the 37th mile from River Trois Pistoles at an elevation of 545 feet above the sea.

The line now enters the valley of the Snellier River, and changing its former course to a southerly direction, it begins to ascend with grades the heaviest of which are 52 and 53 feet to the mile, and together measuring 2.2 miles in length in a distance of about three miles; between the 44th and 45th mile from River Trois Pistoles the line attains an elevation of 786 feet and passes over a water shed to the valley of the North Branch of the Toledi.

Following this Branch of the Toledi in a general southerly direction with undulating grades to the 47th mile, three miles of 64 feet grade are required before Echo Lake is reached at the 50th mile and at an elevation of 985 feet. At Echo Lake the line turns more to the east, and a rapid ascent of 70 feet per mile for three and two tenths miles is unavoidable.

From the 54th mile to the 63rd mile the Railway route will pass at some distance to the east of the surveyed line. At the 56th mile it will reach summit lake 1350 feet above the sea, with grades probably not exceeding 53 feet to the mile, and from the 59th to the 63rd mile, it is believed the grades will undulate easily.

At the 63rd mile the line is 1360 feet above the sea, from this point it follows a tributary of the Rimouski, crosses the Boundary between Canada and New Brunswick at about the 65th mile and then ascends with a grade of 43 feet to Lake Tiarks at the 67th mile, attaining a total elevation of 1515 feet. At this point the line crosses the water shed between the streams flowing into the St. Lawrence and those discharging into the River St. John by the Green River.

From the Lake Tiarks summit, the line passes almost on a level for a mile and a half to the valley of the Green River, and then descends with a grade of 59 feet per mile for nearly two and a half miles, reaching Green River Lake between the 70th and 71st mile. The elevation of this Lake is 1365 feet above tide water.

From Green River Lake the line follows in a south-easterly direction, the valley of the north-west branch of Green River, to the Fork's at the 81st mile. On these ten miles it gradually descends with grades generally less than 30 feet per mile. At the Forks the elevation is 1075 feet.

The line continues in a south-easterly direction from the Forks, ascending gradually the south-east branch of Green River, to a point 82.7 miles from Trois Pistoles, where this section terminates. The elevation here is 1130 feet above the St. Lawrence

The following is an abstract of the grades shown on the profile of the line surveyed on the Trois Pistoles and Green River section :

TOTAL LENGTH IN MILES.

CHARACTER OF GRADES.	Ascending South.	Ascending North.
Grades under 20 feet per mile	16.5	14.1
" from 20 to 30 " "	5.6	9.5
" " 30 to 40 " "	5.1	4.5
" " 40 to 50 " "	7.0	1.7
" " 52.8 " "	4.1	0.8
" " 59.0 " "	0.0	2.4
" " 64 " "	3.2	0.0
" " 70 " "	3.2	0.0

Level.. 5.0 Miles.
Total Length...................................... 82.7 "

There are no rivers of great size on the section above described, and consequently the bridging is comparatively light. The iron bridges required will be of the following general dimensions :

	HEIGHT ABOVE WATER.	No. OF SPANS.	LENGTH OF EACH SPAN.
Over Abawisquash River	22 feet.	1	60 feet.
" Wagan Stream	13 "	1	80 "
" Turadi River	9 "	1	60 "
" 1st Crossing Snellier River	47 "	8	40 "
" 2nd do do	35 "	3	40 "
" 3rd do do	20 "	1	20 "
" 4th do do	20 "	1	20 "
" 5th do do	20 "	1	20 "
" D'Embarras River	9 "	1	50 "
" Toledi do	10 "	1	30 "
" Green do	13 "	1	80 "
" " do	10 "	1	60 "
" " do 3 crossings	12 "	3	60 "

Between the 19th and 71st mile from Trois Pistoles, the line above described makes a very great and objectionable detour to the eastward, which I feel confident can be avoided by a more direct route, and thus save about twenty miles in distance.

From Green River Lake, near the 71st mile running north-westerly, an opening leads through the highlands to the valley of the south-east branch of the River Toledi. The water shed between Green River and the Toledi at this place, is probably not more than fifty feet above Green Lake and here the line can be carried over to the Toledi valley, with a summit about 100 feet lower than the one referred to at Lake Tiarks. After passing the summit, the Toledi must be followed, but this stream falls too rapidly to admit of a Railway being made along the bottom of the ravine, with suitable grades. To make this route available therefore, it would be necessary to descend gradually on the side hill, a plan, which, from the character of the ground, will be somewhat difficult and expensive, and, under any circumstances, long maximum grades will be required.

It was to avoid these difficult and objectionable features that the exploration was carried round by Lake Tiarks. From the accounts of Indians and hunters, there was good reason to expect that a comparatively easy line might be found to the valley of the Abawisquash, without descending to the Toledi and without increasing greatly distance over that by the direct route.

These expectations were however only partially realised, for although the line surveyed has generally very favorable grades, yet its length due to the easterly detour is much too great, and in consequence I would be disposed to recommend the direct route by the Toledi and Sandy Lake. A great deal of careful surveying will be required on this section, before the best and cheapest location can be found along the Toledi, and across from Sandy Lake to Eagle Lake. The work too will prove heavy and expensive; but as twenty miles of Railway will be saved thereby, I am satisfied that the total quantity of work on the whole section, from Trois Pistoles to Green River by the direct route, can scarcely exceed the quantities required to from the circuitous route. And therefore in estimating the probable cost, I shall adopt the quantities computed from the profile of the line surveyed, as those necessary in the building of this section, and of which the following is an abstract :—

1st. Common Excavation.................................	2,391,664 c. yards.
2nd. Assumed proportion of rock excavation.............	90,000 "
Total Excavation.......................	2,481,664
3rd. Culvert Masonry.............................	18,908 c. yards.
4th. Bridge "	7,565 "
5th. Weight of Iron in Bridges.............................	183 Tons.

With the exception of Ballast, which is scarce, it is believed that materials for construction can be procured readily on this section. Stone of different qualities is abundant. Cross-ties will require to be made of the best description of Spruce or Balsam, as other kinds of timber usually employed are rarely met with. With regard to the durability of the Spruce and Balsam found in this district, I am convinced it is fully equal to that of Hemlock, the timber largely employed for cross-ties in western Canada. On the boundary line between New Brunswick and Canada, cut out ten years ago, I saw many trees of the diameter suitable for cross-ties which had lain on the ground during that period, and still to a certain extent sound.

GREEN RIVER FORKS TO RESTIGOUCHE.

Commencing where the last section terminates at an elevation of 1130 feet, the line continues south-easterly about a mile and a half to the mouth of Otter Branch ; it then turns to a southerly direction and ascends a winding valley through a mountainous country to Larry's Lake, the head waters of this branch of Green River; a few hundred yards south Larry's Lake, and near the 7th mile from the beginning of this section, the line passes through the most favorable opening in the highlands that could be found; and here attains a total elevation of 1478 feet, having ascended about 350 feet in seven miles with grades varying from 34 to 70 feet per mile.

The Larry Lake summit divides the waters of Green River from those flowing into the Restigouche, and the line now begins to descend a Tributary of the latter river designated the Gounamitz.

The descent of the Gounamitz is very rapid, involving a continuous grade of 70 feet to the mile for nine and half miles, certainly one of the most unfavorable on the whole line surveyed, but I fear unavoidable. To secure this grade it will be necessary to locate the line along the side hill, which from the character of the ground can be done without much difficulty.

At 16¼ miles from the beginning of this section the elevation is 806 feet, the line from this point continues descending the valley of the Gounamitz to its confluence with the Restigouche near the 32nd mile. The grades for the last 15 miles are remarkably easy, the average about 23 feet to the mile and none exceed 40 feet to the mile. At the end of this secction the elevation of the line is 455 feet above tide water.

The following is an abstract of the Grades shown on the profile:—

	CHARACTER OF GRADES.	TOTAL LENGTH IN MILES.	
		Ascending South.	Ascending North.
Grades under	20 feet per mile.........	0.0	9.0
" from 20 to 30	" "	0.0	5.3
" " 30 to 40	" "	1.0	0.8
" " 40 to 50	" "	2.0	0.0
"	52.8 " "	1.0	0.0
"	61 " "	1.1	0.0
"	70 " "	1.7	9.6

Level............. .. .. 0.8 miles.
Total length of section................. .... 32.3 "

Only three Iron Bridges will be required on this Section, two of which will be over the Gounamitz River. The first in one span of 100 feet and 17 feet above the water. The second in two spans of 80 feet each 14 feet high. The third Bridge will cross the north branch of the Gounamitz, it will consist of two spans each 40 feet and 26 feet above summer water in the river.

Total quantity of the principal items of work on this section as calculated from the approximate profile are estimated as follows:
  1st. Common excavation............................. 1,752,900 c. yards.
  2nd. Assumed proportion of rock excavation...... 66,800 "

                    Total excavation............. 1,819,700 "
  3rd. Culvert masonry..................................... 12,426 "
  4th. Bridge " .... 1,281 "
  5th. Total weight of iron in Bridges .... 130 tons.

Stone suitable for building purposes may be had in the vicinity of the River Restigouche, on the Gounamitz and also on the Green River. Cross-ties may be made of black or grey Spruce of which there is a great abundance, and occasionally Tamarac may be found. Gravel of good quality is everywhere very plentiful on this Section.

RESTIGOUCHE TO TOBIQUE.

After leaving the valley of the Gounamitz, the line runs easterly about a mile and then crosses the River Restigouche at the point where this section begins. The line then ascends the valley of Boston Brook, with grades varying from 50 feet to 70 feet per mile for five and a half miles, when it attains an elevation of 805 feet. At this elevation it continues southerly on a level for a distance of about a mile and a half, then slightly descends to a branch of Jardine's Brook. From Jardine's Brook the line has easy undulating grades along the head waters of Grand River to the 13th mile; it then begins to ascend through fine hard-wood land with grades of 65 feet per mile to the middle of the 18th mile, where it reaches an elevation of 1074 feet. The line now descends with favorable grades to Salmon River, which it crosses at the 23rd mile at an elevation of 858 feet. At the 30th mile after crossing various branches of Cedar Brook on easy undulating grades, it passes at an elevation of 830 feet, over a summit between a tributary of that stream and Two Brooks. It then follows Two Brooks on descending grades, chiefly under 40 feet to the mile, to the north bank of the River Tobique, which it reaches at the 39th mile and at an elevation of 445 feet above the sea; continuing in a southerly direction along the north bank of the Tobique, on almost level grades, the line reaches a favorable point for crossing near the mouth of the Little Gulquac, where this section terminates at a total distance of 45.4 miles from the Restigouche.

The following abstract will show the character of the grades on the section above the described.

				TOTAL LENGTH IN MILES.	
CHARACTER OF GRADES.				Assending South.	Ascending North.
Grades under 20 feet per mile				2.0	4.2
"	20 to 30	"	"	0.6	2.9
"	30 to 40	"	"	1.1	5.9
"	40 to 50	"	"	1.6	0.6
"	50 to 52	"	"	2.1	2.1
"	54	"	"	0.0	1.3
"	60	"	"	1.0	5.7
"	65	"	"	6.8	0.0
"	70	"	"	1.0	0.0

Level.. 6.5 Miles.
Total length of Section............................... 45.4 "

The Bridging required on this section consists, firstly, of one across the River Restigouche, about fifteen feet above the water and in five spans of 60 feet each; secondly, of a Bridge 25 feet high with two sixty feet spans across the Salmon river; thirdly, of one across the River Tobique having three spans 100 feet each, and about 32 feet above summer water; arch and beam culverts will suffice for all other waters crossed.

The quantity of Excavation and other work on this section has been calculated from the approximate profile and the following is presented as an abstract:

1st Common Excavation............................... 2,068,600 c. yards.
2nd Assumed proportion of Rock Excavation......... 456,500 "

Total Excavation................... 2,525,100
3rd. Culvert Masonry .... 13,787 "
4th. Bridge " .... 1,469 "
5th. Weight of Iron in Bridges......................... 276 Tons.

Good stone for constructing the Restigouche and Tobique bridges may be had at no great distance from the bridge sites; materials for the construction of culverts within ten miles of both rivers may also be obtained without much difficulty, but on the intermediate parts of the line it has not been ascertained that stone can be procured. Sand is plentiful and it is believed that gravel will be fund upon or close to the line. Tamarac as well as spruce cross-ties, can be had in the district passed through from the Restigouche to the Tobique Rivers.

TOBIQUE TO KEDSWICK SUMMIT.

This section commences at the River Tobique near the mouth of the Little Gulquac; a position which was selected for crossing the Tobique, in the expectation that the surveying party would intersect a line cut out by the Capt. Henderson towards the Miramichi, and thus save time and expense in carrying on the examination through part of this section No advantage was gained by this stop, as the old line was so entirely obliterated in many places, that it could only be traced with the greatest difficulty, and in consequence it was found expedient to abandon the old survey and to take an independent course. The line commences at an elevation of 425 feet, and ascends the valley of the Little Gulquac, with grades varying from 36 to 63 feet per mile for five miles; it then passes over a ridge to the Little Wapsky River and continues on easy grades to the end of the 11th mile. The line now crosses the Wapskykegan, where a bridge of great magnitude will be required, and begins to ascends on a maximum grade of 70 feet per mile to a summit at the head of Oven Rock Brook. The summit is reached at 16¾ miles, and the elevation attained is 1170 feet above the sea. Between the River Wapskyhegan and the summit, the greatest difficulties on this section are found. Besides the Wapskyhegan bridge, which will be nearly a thousand feet long and 140 feet high, the excavation on this ascent, five and a half miles long, will be unusually heavy.

The line then enters, by Franks's Brook, the valley of the north branch of the Miramichi, which it follows, crossing the river twice near the 22nd and 23rd miles. From the 23rd mile to the 32nd, the line winds along the west bank of the river; then strikes across a Cariboo plain to the north-west branch of the Miramichi, which it reaches at the end of the 37th mile, with an elevation of 783 feet above the sea. The grades are all descending from the summit to the north-west branch, and are remarkably easy, being generally on this distance of 21 miles under twenty feet to the mile, and only in one instance as high as 44 feet to the mile.

Crossing the north-west branch of the River Miramichi, about a mile westerly from the "Forks," the line ascends by Turtle-shell Brook, without difficulty to the water-shed between the last named river and the Nashwaak, which it reaches at the beginning of the 40th mile at an elevation of 950 feet. Descending on a favorable grade for about a mile, the line then follows the River Nashwaak on the westerly side, and on nearly level grades to the 51st mile, where the Two Sister Brooks fall into the main stream. At this point, the Nashwaak leaves the southerly direction which it previously maintained, and turns nearly at right angles to the east. The line, however continues southerly, and ascending by one of the Two Sisters, reaches the Keswick summit at about the 54th mile, and at this point attains a height above the tide of 1005 feet. From the summit the line descends on a 65 feet grade for a distance of about a mile, to a point a little easterly from Lake Beccaguimic, where this section of the survey terminates.

The following is a general abstract of the grades taken from the profile of the line surveyed from the River Tobique to the point last referred to:—

TOBIQUE TO KESWICK SUMMIT—CHARACTER OF GRADES.	TOTAL LENGTH IN MILES. Ascending South.	Ascending North.
Grades under 20 feet per mile	1.5	13.6
" 20 to 30 " "	1.7	1.1
" 30 to 40 " "	2.9	5.7
" 40 to 50 " "	0.7	3.1
" 52.8 " "	2.2	0.0
" 56 " "	1.5	0.0
" 63 " "	0.9	0.0
" 65 " "	0.0	1.4
" 66 " "	0.0	0.1
" 68 " "	2.7	0.0
" 69 " "	1.7	0.0
" 70 " "	5.6	0.0
Level	8.3 miles.	
Total length of section	55.6 "	

The Bridging on this section will be heavier than on any of the others. The Little Wapsky will require a viaduct about 55 feet high, and the one across the Wapskyhegan will be 142 feet above the level of the River. The former is proposed to consist of sixteen girder spans, each sixty feet, and the latter of three 100 feet spans over the Wapskyhegan River with 13 sixty feet spans in the approaches. Between the 22nd and 23rd mile, the north-west Branch of the Miramichi will be bridged twice with sixty feet single openings, the one will be 25 feet high, and the other 18 feet. A fifth bridge will be required over the south-west branch 20 feet in height, and it is proposed to adopt three spans for this work, the centre span one hundred feet, the other two each 60 feet.

The quantities calculated from the profile deduced from the survey of this section of the line are as follows :—

1st. Common Excavation .... 2,266,700 cubic yards.
2nd. Assumed proportion of Rock Excavation 336,400   "

Total Excavation 2,603,100

3rd. Culvert masonry 19,320
4th. Bridge   "   13,500
5th. Weight of Iron in Bridges............ .... 794 Tons.

Good stone for Bridge masonry can be had on and near the River Tobique, and sandstone suitable for the same purpose can be obtained on the Miramichi and Nashwaak Rivers; stone for culvert masonry may be obtained without much difficulty throughout the section. There is also good sand for building purposes, and abundance of gravel for Ballast.

The timber available for Cross-ties, between the River Tobique and Keswick Summit, consists of Spruce, Tamarac, Hemlock and Cedar.

KESWICK SUMMIT TO LITTLE RIVER.

The line enters the Keswick valley near the source of the west branch, and continues within its limits until the River St. John is reached; the descent of the west branch is very rapid for the first eight or nine miles, and heavy grades for this distance will be unavoidable. The maximum grades shown on the approximate profile of this section are 66 feet to the mile, and to obtain this on the line by the west branch, heavy side hill work will be necessary for a considerable distance.

Probably the east branch may offer a more favorable approach to the main valley of the Keswick River. But the season was too far advanced to admit of a proper examination by this route being made.

From the ninth mile the line winds along the side of the River, occasionally crosses it, and then continues on the flats until it finally reaches the north side of the River St. John at the 29th mile. For twenty miles, up to this point, the grades are remarkably favorable, in no case being over 40 feet to the mile and generally under 20 feet to the mile.

From the mouth of the Keswick the line runs along the north bank of the River St. John almost on a dead level, crossing the River Nashwaaksis at the 37th mile. It reaches the Fredericton upper ferry at 38½ miles, and the lower ferry at the end of the 39th mile; about three-quarters of a mile farther on the line arrives at the Nashwaak, an important river, 500 feet in width where it is crossed.

Soon after crossing the Nashwaak, the line leaves the banks of the St. John, and, turning round Barkers hill, follows an easterly direction with very favorable undulating grades to the Little River, where this section of the survey terminates.

The following table is an abstract of the grades shown on the profile:—

CHARACTER OF GRADES FROM KESWICK SUMMIT TO LITTLE RIVER.	TOTAL LENGTH IN MILES.	
	Ascending South.	Ascending North.
Grades under 20 feet per mile...........................	11.0	10.4
" 20 to 30 " "	1.0	3.1
" 30 to 40 " "	0.0	3.8
" 40 to 50 " "	1.7	0.0
" 52.8 " "	0.0	1.0
" 66 " "	0.0	3.0
Level	16.6 miles.	
Total length of section	61.6 "	

With the exception of the Nashwaak, the rivers to be crossed on this section are unimportant. The spans given in the following list will probably be sufficient.

	Height.	No. of Spans.	Length of Spans.
Over North-West branch of Keswick............	20 feet	2	50 feet.
" North-East do do	12 "	1	75 "
" Little Fork's River	18 "	1	50 "
" Nashwaaksis River.............................	18 "	1	75 "
" Nashwaak.......	20 "	7	75 "
" Noonan's Brook.................................	14 "	1	30 "
" Burpee's Brook....	13 "	2	50 "

The approximate profile made from the survey of this section shows that the following quantities of the chief kinds of work are sufficient:—

1. Common Excavation..1,904,100 C. yds.
2. Assumed proportion of Rock Excavation.....................   170,000   "

    Total Excavation..2,074,100   "
3. Culvert Masonry..   14,931   "
4. Bridge   do   3,410   "
5. Iron in Bridges...   320 tons.

There will probably be some difficulty in procuring building stone, at least for the Bridge Masonry, within a convenient distance along the Keswick valley, as none suitable appeared to crop out along the line of survey; fortunately, however, the bridging in this quarter is comparatively light. From the Keswick to the Little River it is believed that stone for all the bridges and culverts may be found readily. Material for ballast, although not of the best quality, can be had in abundance on this section. The timber for cross-ties, in this locality, consists of Spruce, Hemlock and Cedar.

LITTLE RIVER TO COAL CREEK.

From Little River the line continues in an easterly direction to the head of the Grand Lake Navigation, on the Salmon River, which it crosses at the 19th mile. For this distance the grades are undulating and favorable; near the 9th mile the line crosses the Newcastle River, and in this locality it passes close to several coal mines, where coal, of fair quality, crops out on the surface; at the 16th mile the line crosses an arm of "Iron Bound Cove" which will have to be bridged.

After passing Salmon River the line curves southerly, and passes over a ridge with ascending and descending grades of about 60 feet per mile, to Coal Creek, which it reaches near the 25th mile; about a mile and a half farther south, the line joins on to the next section.

The profile shows the following grades:

CHARACTER OF GRADES FROM LITTLE RIVER TO COAL CREEK.	Ascending South.	Ascending North.
Grades under 20 feet per mile............................	1.5	2.0
" 20 to 30 " "	0.0	1.1
" 40 to 50 " "	0.0	3.0
" 52.8 " "	3.9	0.0
" 58 " "	0.0	1.6
" 60 " "	0.0	1.0
" 61 " "	2.2	0.0
" 65 " "	1.9	0.0
Level...	8.1 miles.	
Total length of Section..............................	26.3 "	

The bridging on this section is very heavy when its length is considered. The rivers to be crossed and the structures proposed are as follows: of course the character and dimensions of the latter may be greatly modified on a proper location survey being made.

At Little River the bridge will be 45 feet in height with nine spans, one of 100 feet and eight of 60 feet openings.

At the Newcastle River the bridge will be 37 feet high and will have eight spans, one 100 feet and seven of 60 feet openings.

At Iron Bound Cove the bridge will be 23 feet above the level of the water, and it will have three spans each 60 feet.

At Salmon River it is proposed to have a bridge 17 feet in height with nine spans each 60 feet.

At Coal Creek a viaduct of considerable magnitude is at present considered necessary; the height will be about 70 feet, with one span of 100 feet across the stream and eleven 60 feet spans in the approaches.

The calculation of quantities from the profile of this section gives the following totals:

   1st. Common Excavation................................... 734,125 C. Yds.
   2nd. Culvert Masonry..   6,297   "
   3rd. Bridge   10,683   "
   4th. Bridge Iron...   834 Tons.

The most convenient point for obtaining building stone has not been ascertained.

But as the proposed bridges are either on or within a short distance of Grand Lake, which is navigated by steamboats running to St. John and Fredericton, it is thought that the supply of building material will not be difficult, even should the immediate locality not produce it.

Gravel for Ballast is plentiful. The timber for Ties produced in this district is Spruce, Tamarac, and Prince's Pine.

COAL CREEK TO APOHAQUI.

After ascending from Coal Creek with a 65 feet grade, the line follows a southerly direction over a favorable country, and reaches Canaan River near the eleventh mile.

Canaan River is crossed at Long Rapids, and the line there ascends by Porcupine Brook, on grades generally 60 feet per mile to Long's Creek Bridge, which it reaches at the 15th mile. The line then decends to the North Branch of Long's Creek, which it crosses at about the 17th mile; then continues in a general southerly direction up the valley of the South Branch, on grades not exceeding 52.8 feet per mile; it passes over a ridge and enters Chowan's Gulch, a little beyond the 21st mile.

Chowan's Gulch leads the line by a rapid descent, involving grades of 52.8 and 60 feet per mile, for five and a half miles, to the valley of Studholme Mill Stream; following which on undulating grades to about 31½ miles, it joins the European and North American Railway at Apohaqui Station.

The following is an abstract of all the grades on this section.

CHARACTER OF GRADES FROM COAL CREEK TO APOHAQUI.	TOTAL LENGTH IN MILES.	
	Ascending South.	Ascending North.
Grades under 20 feet per mile	1.0	0.5
" 20 to 30 " "	0.4	1.9
" 30 to 40 " "	1.3	1.1
" 40 to 50 " "	1.3	0.4
" 52.8 " "	4·7	6·9
" 60 " "	3.7	4.2
" 65 " "	0.8	0.0
Level	3.4 miles.	
Total length of Section	31.6 "	

The bridge over the Canaan River will be the most costly structure on this section, its height above the water will be 55 feet, and it is proposed to have six openings, one in the centre of 150 feet span, and five others each 60 feet span.

The next bridge will be over the north branch of Long's Brook, it is intended to have three thirty feet spans, its height will be nearly thirty feet.

Sharp's Brook, about the middle of the 29th mile, will require to have a single span bridge of 40 feet, and 21 feet high.

The last bridge on this section will be over the Kenebeccasis River, about 400 yards from Apohaqui Station, it will be 21 feet above summer water, and will have five spans, a centre one 150 feet in length, and four others each fifty feet long.

The approximate quantities of work on this section are as follows:—

1st. Common Excavation	850,860	cubic yards.
2nd. Assumed proportion of Rock Excavation	216,360	"
Total excavation	1,067,220	
3rd. Culvert Masonry	18,040	"
4th. Bridge Masonry	4,170	"
5th. Bridge Iron	386	Tons.

It is reported that the locality around Canaan River and Porcupine Brook will afford good stone for heavy masonry. A sandstone crops out at other points along this section, but it is not sufficiently exposed to enable one to judge of its quality. Stone for culvert masonry in all probability can be had without much difficulty. There will be no difficulty in obtaining good gravel for Ballast.

On this section Tamarac is abundant, and most of the other descriptions of Tie-Timber already mentioned can be had.

In concluding these remarks on the character of the line surveyed through the centre of New Brunswick, I may allude briefly to its leading features.

The course taken by the line above described from the River du Loup towards the southern part of New Brunswick is generally direct and at some distance from the eastern Frontier of Maine. Except at one point, this distance is not less than that between the Grand Trunk Railway east of Quebec, and the northern boundary of the same state; the point referred to lies to the north and east of Grand Falls on the River St. John. I may mention, however, that at this point, which lies between the Restigouche and the Tobique, I instituted a supplementary exploration after the survey was finished and the discovery was made that the line approached the Frontier nearer than desired. This exploration resulted in showing, that there is every probability of a favorable location being obtainable, without keeping so close to the Boundary of the Province at this point. The alternative line, which possibly can be had between the Restigouche and Tobique Rivers, is shown on the general map of the country which accompanies this.

The line continues on a course towards the city of St. John, generally direct until Fredericton is reached. From Fredericton it was my object to find the shortest route to St. John on the east side of the river, the crossing of which is, in some respects, objectionable.

To reach St. John on the easterly side of the river it was found necessary, on account of difficulties that could not be easily overcome, to pass round by the head of Grand Lake; and in this direction, though rather circuitous, a favorable line was found to a point of connection at Apohaqui with the existing railway leading to St. John. This is probably the most direct line that can be had to the City of St. John from Fredericton, without crossing the river.

By crossing the river in the neighborhood of Fredericton, St. John may be reached much more directly by way of Oromocto and Douglas Valley, on a line carefully surveyed last summer by Mr. Burpee for the New Brunswick Government, copies of the plans of which have been placed in my possession. This would, without question, be the most direct central route from Canada to the Harbour of St. John on the Atlantic seaboard. The distances by the several projected lines will be particularly referred to hereafter.

The following general abstract will give an idea of the grades which may be expected on the whole length of the surveyed line beginning at River du Loup and ending at Apohaqui Station:

CHARACTER OF GRADES ON WHOLE SURVEYED LINE FROM RIVER DU LOUP TO APOHAQUI.	TOTAL LENGTH IN MILES.	
	Ascending South.	Ascending North.
Grades under 20 feet per mile	41.9	58.4
" from 20 to 30 " "	9.6	29.9
" 30 to 40 " "	13.9	21.8
" 40 to 50 " "	15.2	9.8
" 51.9 " "	2.1	2.1
" 52.8 " "	18.3	8.7
" 53.5 " "	....	1.1
" 54 " "	....	1.3
" 56 " "	1.5	....
" 58 " "	....	1.6
" 59 " "	....	2.4
" 60 " "	4.7	10.9
" 61 " "	3.3	....
" 63 " "	0.9	....
" 64 " "	3.2	....
" 65 " "	9.5	1.4
" 66 " "	....	1.0
" 68 " "	2.7	8.0
" 69 " "	1.7	....
" 70 " "	11.5	9.6
Level	51.7 miles.	
Total length	360.0 "	

The above are the actual grades on the profile of the line surveyed, but as the direct route from Eagle Lake to Green River, referred to in the foregoing, will cut off a portion of the above line, a certain alteration in the table of probable Grades will be necessary. The direct route between these points has not been instrumentally surveyed, and therefore the precise character of the grades is not known. It is believed, however, that whilst the construction of the Railway on the direct route from Eagle Lake to Green River would shorten the distance 20 miles, and thus reduce the whole length of line to 340 miles, it would, at the same time, involve the adoption of a long ascending grade of a heavy character, from near Sandy Lake, in the valley of the Toledi, to a summit near the Canada and New Brunswick Boundary Line.

Without doubt, some of the grades shown in the Table are severe. But perhaps they are not more so than could reasonably be expected, when the peculiar character of the country, crossed by this line, is taken into consideration; a maximum grade of 70 feet per mile is not greater than the maximum on the Railway from Truro to Halifax, which must form a portion of the whole line between the latter city and Canada. Nor is it greater, as I am informed, than the maximum on the Portland Division of the Grand Trunk Railway. The ascents, however, on the line surveyed, if not steeper, are much longer where they do occur than those on either of the two railways named.

It is, perhaps, fortunate that the unfavorable grades are confined to particular points, instead of occurring at frequent intervals throughout the whole extent of the line; as, in the event of this line being selected and constructed, it could be worked with greater advantage and economy, by employing extra engine power on heavy trains, only at those points, instead of being obliged to use it throughout. It would be impossible to economize engine power, and thus prevent unnecessary wear and tear, on level sections of the line, were the maximum grades distributed.

It happens that there are, in all, four points where gradients of an unfavorable character occur, two of which are ascending south and two ascending north.

The two where the gradients ascend south, are situated at the head of the Toledi and at the Wapskyhegan. The Toledi gradient is about 70 miles from the River-du-Loup, and the Wapskyhegan ascent is about 100 miles still farther south.

The two gradients ascending north are about 125 miles apart, one is situated at the head of the Keswick valley, and the other at the head of the Gounamitz valley.

If the length of the ascents at these four points be deducted from the length of the whole line, it will be found that 48 per cent of the remainder is level, or under 20 feet to the mile; thirteen per cent., from 20 to 30 feet per mile; eleven per cent., from 30 to 40 feet per mile; eight per cent., from 40 to 50 feet per mile; nine per cent., 52.8 per mile; seven per cent., from 52.8 to 60 feet per mile, and four per cent., from 60 to 66 feet per mile.

In concluding the description of the main features of the line surveyed through the centre of New Brunswick, I desire to add that the survey can scarcely be considered much more than a mere exploration. The impenetrable character of the forest, more particularly to the north of the River Réstigouche, the difficulties experienced in getting supplies forwarded through the woods, together with the limited time and means allowed for the service, rendered it impossible to accomplish more than a rough and rapid instrumental survey of a line, in all probability not the best than can be found through the country. However, sufficient information, it is hoped, has been procured to show, not only that a practicable line can be obtained, but also (although no great accuracy is professed) what it may possibly cost.

Plans of this survey have been made on a scale of 500 feet to an inch horizontal. On these plans the line chained and levelled over is distinct from the railway line, the latter is shown in red, with regular curves and tangents, and it runs in the direction which it is thought a trial might take. Deviations from this line would no doubt be found necessary at many points, on more exact surveys being proceeded with; but it is believed that although the alignment may frequently be changed, yet neither the gradients nor the work need necessarily be increased.

The approximate profile is intended to represent the probable surface of the ground, the gradients, the cuttings, embankments, and other work on the "Railway line;" it is compiled from the measurements and levels taken on the Survey line, that is, the line cut out through the woods, and also from such cross sections or lateral explorations as were made or deemed necessary. Where the "Railway line" is on, or near the line levelled over, the profile may be considered correct; where these lines are some distance apart the former must be received as approximate only.

The quantities of work herein submitted are calculated from the approximate profile above referred to and, as far as known, are correct and ample.

All the through cuttings are estimated to be 30 feet in width at formation level. Side cuttings 24 feet wide, and embankments 18 feet wide.

Openings over 20 feet in width are estimated to be wrought Iron Tubes or Girders resting on substantial masonry. All openings under twenty feet are estimated to be Arch or open Beam Culverts.

The following are the total quantities of the chief classes of work, calculated as above described, and considered sufficient to complete the Bridging and Grading of the line, in a permanent and substantial manner, from the River du Loup to Apohaqui, a distance of 340 miles.

Total Excavation	13,828,923	cubic yards.
Assumed proportion of common Excavation	12,453,238	" "
Assumed proportion of Rock Excavation	1,375,695	" "
Culvert Masonry	107,725	" "
Bridge do	49,039	" "
Bridge Iron	3,337	Tons.

THE MATAPEDIA SURVEY,

Lest the explorations through the centre of New Brunswick should prove unsuccessful, and the route by Bay Chaleurs, recommended by Major Robinson in 1848, should under any circumstances appear entitled to the preference, I deemed it expedient to have a careful examination made of the section which that gentleman as well as Captain Henderson considered the most difficult and expensive between Halifax and Quebec.

"The most formidable point of the line is next to be mentioned—this is the passage up the Matapedia valley.

"The hills on both sides are high and steep and come down either on one side or the other pretty close to the river's bank and involve the necessity (in order to avoid curves of very small radius) of changing frequently from one side to the other. The rock too is slaty and hard; from this cause 20 miles of this valley will prove expensive but the grades will be very easy.

"About fourteen bridges of an average length of 120 to 150 yards will be required up this valley. There is also a bridge of 2,000 feet long mentioned in the detailed report as necessary to cross the Miramichi River. *Report of Major Robinson, 31st August,* 1848."

"The section of country lying between the Restigouche and St. Lawrence rivers is a vast tract of high land, intersected in every direction by deep valleys and vast ravines through which the rivers flowing to the St. Lawrence and Restigouche wind their course.

"The height of land from which those rivers flow respectively north and south is full of lakes and along them the mountain ranges rise to a great elevation.

"The average distance between these two Rivers is about 100 miles.

"The only available valley which my knowledge of the country, or the explorations we have carried on enable me to report upon, by which a line of Railway can be carried through this mass of high lands is that of the Matapediac River.

"This valley extends from the Restigouche to the Great Matapediac Lake, a distance of between 60 and 70 miles, and as the summit level to be attained in this distance is only 763 feet above tide water, the gradients, generally speaking, are extremely favorable.

"From the broken and rocky character of this section of country some portions of this part of the line will be expensive, especially the first 20 miles of the ascent, in which the hills in many places come out boldly to the river, and will render it necessary to cross it in several places.

"The rock formation is nearly all slate; there are settlements on the Matapediac River, as far as the mill stream.

"Generally speaking, however, the greater portion of this section of country is unfit for cultivation, consisting of a gravelly rocky soil covered with an endless forest of spruce, birch, pine, cedar, &c.

"From the mouth of the river as far a the 365th mile the line continues upon the east bank; above this, at the mouth of Clark's Brook the rocky bank of the river is very unfavorable, and to obtain proper curves it crosses to the point opposite and then recrosses immediately above to the more favorable ground on the east bank, between this and the mouth of the Ammetssquagan River, the line to obtain good curves and avoid those places where the hills come out bold and rocky, crosses the river four times.

"The position of the line for three miles above and below Ammetssquagan River, where the hills are steep and rocky close to the River, will be the most expensive part of the line.

"Above this the line follows the eastern bank to the 377th mile. The hills on either side are very high, but the eastern bank is pretty favorable; between the 378th and 380th mile the river turns twice almost at right angles. Shut in on the south by a rocky precipice 150 feet high.

"It will be necessary to cross the river three times here. The centre bridge will be a heavy one; but there is an island at the elbow which will serve as a natural pier. Above this, from the 380th mile to the Forks (the mouth of the Casupsent River,) at the 395th mile, the valley becomes more favorable. The hills on either side are not so lofty and recede farther from the river, the line crosses the river twice between the 385th and 390th mile to avoid a rocky precipice on the left bank; and again about one mile below the Forks, making in the first 38 miles, up the valley of the Matepediac, twelve bridges in all. These bridges will average from 120 to 150 yards long. *Report of Captain Henderson,* 1848."

The object of the examination was to ascertain the exact nature of the difficulties alluded to, if they could be more cheaply overcome or avoided, and also with a view to form an estimate of the whole expenditure required to construct this section. With this data the cost of the whole line it was supposed could be ascertained with sufficient accuracy, by adopting an ordinary average charge per mile for the remainder of the line, which the gentlemen referred to reported as extremely favorable and easy of construction.

With this view I instituted a thorough survey of the Matapedia river and valley, beginning at its junction with the Restigouche and running northerly. The Transit, Chain, and Level were used throughout. A longitudinal section was made from the Restigouche to the waters of the St. Lawrence, and cross sections were also made, whenever it appeared necessary, to ascertain the character of the slopes of the adjacent ground. The survey was continued northerly until the waters leading to the St. Lawrence were reached. The field work is laid down to a scale of 200 feet to one inch, on the plans which accompany this; and should the Matapedia route ever be selected, the carefully prepared plans and other information derived from this survey, will be found of such a character, as will enable the location of the line to be proceeded with, for a distance of about 70 miles, without additional preliminary examinations of any consequence.

I shall now proceed to describe briefly the Engineering features of the line surveyed.

The River Matepedia flows in a direction from north-west to south-east, it takes its rise within twenty miles of the banks of the St. Lawrence, at Grand Métis, and discharges into the River Restigouche some 16 miles west of the Port of Cambeltown. From the point where the River Causapscal joins the Matapedia, known as "The Forks," to the Restigouche, a distance of 35 miles, the river flows through a rocky gorge with many twists and windings, between banks on both sides, varying from 500 to 800 feet in height These banks are in many places very precipitous, and rise immediately from the river's edge, but frequently there is a narrow flat margin, favorably situated for a road or railway Above the Forks the character of the country is different, the high banks begin to recede from the river, and although frequently rough ground is encountered, there are no obstacles of much consequence.

The best point for bridging the River Restigouche, is still a question for future consideration. The line surveyed follows the easterly side of the Matapedia, and therefore in the event of this location being adopted, the bridge over the Restigouche would necessarily be placed below the junction of the two rivers; for a certain distance at least, the line would have an equally good location to the west of the Matapedia, and there would be some advantage, in crossing the main river, above the point where the Matapedia discharges into it. Although this is an important question of detail, it need not now be further alluded to.

The section to be described, of which an approximate profile is prepared, and quantities calculated, is 70 miles in length, and the miles are numbered on the plan from the north to the south. It will be more convenient, however, to describe the features of the line, beginning at the Restigouche, and running northerly. The 70th mile ends immediately opposite the farm house of Mr. Daniel Fraser, on the flats where the Matapedia joins the Restigouche.

At seven miles from the mouth of the Matapedia, Clark's Brook is crossed. Up to this point the general course of the river is straight, and a direct line can be had without much curvature, and with remarkably easy grades. The sharpest curve on this distance is a short 4° curve (1432 feet radius) below Noonan's Gulch, and the heaviest grade is 38 feet to the mile.

At Clark's Brook the River takes a great bend to the west, necessitating a long curve of 1763 feet radius. At the 62nd mile the river again bends to the north, involving a compound curve with radii varying from 1430 feet to 3830 feet. From this point up to "Hell's Gate," about the 58th mile the curvature is easy, although frequent. Immediately north of Hell's Gate a sharp point of rock has been cut through, and the Asmaguagan River, a tributary of the Matapedia, is then crossed.

From the Asmaguagan, the line winds along the easterly bank of the Matapedia, with almost level grades to Connor's Brook, between the 53rd and 56th mile; where ascending and descending grades of 52.8 and 50 feet per mile, for about half a mile, are required to avoid a sharp curve.

About two miles farther up at a place called "the Lewis Rocks" the river takes several sudden twists, and it will be necessary either to form a tunnel through the Lewis Rocks 1300 feet long, or divert the river; the latter would prove the cheapest and is recommended. Above this point for about the third of a mile, the channel of the river will require again to be changed. The works of excavation for about a mile in length in the neighborhood of the Lewis rocks will be unusually heavy.

From the 51st to the 40th mile, the general course of the river is straight, and the line continues along its easterly side with favorable grades and easy curves.

At the 40th mile the line leaves the edge of the river for about two miles, and striking across a point of low land avoids a short bend at the outlet of Metallics Brook.

The next difficulty occurs near the 36th mile where the river takes two exceedingly sharp turns, first easterly, then northerly, at points about three quarters of a mile apart. Fortunately at the first turn, designated "the Devil's Elbow," a piece of low ground at the base of the hills admits of a curve of 1910 feet radius. At the second turn, known as "Alick's Elbow," it will be necessary to throw the line into the river and across an island on a curve of 1430 feet radius. The channel for the river, to the west of the Island, being at the same time increased in width.

The forks of the Matapedia are near the 35th mile; at this point the river is crossed, and the line afterwards follows its westerly bank to the Little Lake, which it reaches at the 30th mile.

Proceeding northward with favorable grades and curves, the line crosses the river Amque at the 22nd mile, and arrives at the Matapedia Lake a mile farther on.

Continuing northerly along the westerly side of the Lake, with the exception of one long curve of 1763 feet radius, near the 17th mile, rendered necessary in order to avoid a high ridge, the line is extremely favorable up to Pierre Brochu's, at the 8th mile; the curves on this distance being in general 5730 feet radius.

At Pierre Brochu's the line leaves the Lake, crosses the Sayebec River at the 7th mile, and ascends by a long grade, part of which is 60 feet to the mile, to the summit Lake, about the middle of the 3rd mile. This is the only instance of a 60 feet gradient, up to this point, from the mouth of the Matepedia.

At the 2nd mile, the water shed between the Restigouche and St. Lawrence is reached, and the elevation at this point above the sea is 794 feet. The line now begins to descend towards the St. Lawrence by the River Blanche, a branch of the Tartigan, and in two

6

miles it reaches the beginning of the northerly end of the seventy mile section, which has just been described.

From the point last mentioned, the survey is carried on by the valley of the River Tartigan, and a line can be had along this river with only an occasional difficulty. The Tartigan flows in a narrow and rather crooked valley, necessitating frequent crossings or deviations of the river, and sometimes a heavy excavation through a projecting point of land; it continues westerly for about six miles, and then turns to the north; up to this point a favorable line can be had. From this point a line was cut and levelled to the Metis River, by Paquett's Brook, but the result was not satisfactory.

Between the River Tartigan and the Metis, a distance of about 14 miles, the country is very broken and irregular in its features, high ridges with deep gulches between are constantly met with. The Metis itself lies in a low wide valley, and it must either be crossed at a high level. on a viaduct of formidable dimensions, or a line must be found by which a favorable descent to the valley can be had. The latter has not been discovered, although from personal explorations I am led to believe that there is a reasonable chance of one being found A great deal of time will yet require to be spent in this locality, in thoroughly surveying the country, before the best line from the Tartigan to the Neigette River, across the Metis Valley, can be determined.

Although the chaining and levels were carried through to St. Flavia, on the shore of the St. Lawrence, a total distance of nearly 100 miles, the line surveyed may be said to terminate at 70 miles north from the Restigouche; from thence northerly the country is only imperfectly explored.

The difficulties met with in crossing the Metis Valley, were not anticipated, as they are scarcely alluded to in the reports on the survey made in 1848. Yet my present impression is that they are perhaps the most serious on the Bay Chaleurs route. Further surveys may however modify this view.

I regret exceedingly, that circumstances would not justify me in incurring the expense of continuing the survey to a more satisfactory issue in this quarter.

I may now, to illustrate more particularly the character of the line surveyed, from the Restigouche, to the point where the water shed between that river and the St. Lawrence is crossed, and the valley of the Tartigan reached, present an abstract of the curves and grades on this section, 70 miles in length.

CHARACTER OF GRADES —MATAPEDIA SECTION.	TOTAL LENGTH IN MILES.	
	Ascending South.	Ascending North.
Grades under 20 feet per mile	6.9	11.7
" 20 to 30 " "	4.4	9.2
" 30 to 40 " "	2.4	7 0
" 40 to 50 " "	1.6	2.8
" 50 to 52 8 " "	2.8	6 2
" 60 " "	0.0	2.7
Level	12.3 miles.	
Total length of Section	70	"

The wrought iron bridging on this section will be as follows, all the other openings are intended to have either arch or beam culverts.

    1st Over River Blanche on 1st mile one span of 50 feet,
    2nd " Savabee River on 7th " 3 spans 50 "
    3rd " River St. Pierre on 9th " 1 span 60 "
    4th " " Tobigote on 19th " 1 " 50 "
    5th " " Amqui on 23rd " 3 spans 60 "
    6th " Indian Brook on 25th " 3 " 40 "
    7th " River Matapedia 36th " 1 span 150 "
    8th " " Assmaguagau 58th " 1 " 80 "
    9th " Clark's Brook 64th " 3 spans 80 "

Whilst the grades are favorable, and the bridging light, it might naturally be expected that the curvature would be excessive, when the tortuous character of the River Matapedia, more particularly below the Forks, is taken into consideration. The following abstract will show, however, that sharp curves have been avoided. The minimum radius adopted on the Grand Trunk Railway (Portland Division), namely, 1,146 feet, not being reached.

CURVATURE.

1° or 5730 feet radius total length			6.1	miles.
1¼° " 3820	"	"	6.9	"
1¾ " 3274	"	"	0.3	"
2 " 2865	"	"	8.9	"
2½ " 2292	"	"	0.1	"
3 " 1910	"	"	6.1	"
3¼ " 1763	"	"	1.8	"
3½ " 1637	"	"	2.6	"
4 " 1432	"	"	3.0	"
4¼ " 1348	"	"	0.3	"
4½ " 1273	"	"	0.6	"
Tangents............		....	33.3	"

Total length of section .... 70.0 "

In submitting an estimate of the quantities of the chief classes of work required to complete the Bridging and Grading on this section, it may be remarked that although the survey and the calculations have been made with great care, I have deemed it prudent to add ten per cent. to all the quantities, to cover any possible oversight, or contingency, connected with the works of construction on this section.

Approximate quantities.

1st. Common Excavation........................1,408,936 Cubic yards.
2nd. Rock Excavation, assumed proportion........... 190,905 "

Total excavation............1,599,841 "
3rd. Culvert masonry...... ....;. 29,317 "
4th. Bridge   do   .... 4,535 "
5th. Iron in bridges .... 350 Tons.
6th. Slope walling to protect embankments on rivers, 63,030 C. yds.

With regard to building materials; the rock exposed along the river is chiefly slate, and although some of it may suit for culverts and slope walling, it would not answer for heavy Masonry. About three miles below " The Forks" I am informed that extensive beds of Sandstone, suitable for Bridge Masonry, can be found. From " The Forks" northerly to the River Amqui, a distance of about 12 or 18 miles, there are few exposures, and the rock where seen is dark shale. From the Amqui, northerly, along the side of Lake Matapedia, a few exposures of Limestone and white Sandstone are seen; the former is not considered of good quality for Bridge Masonry, but the latter is suitable for all kinds of work.

From Lake Matapedia to Metis Valley, the rocks met with are Limestones, Conglomerates, red and grey Shales, and red and blue Slates.

Abundance of Material for Ballast can be had, indeed many of the embankments will consist of nothing else.

Tamarac, Spruce and Cedar will be available for Cross-ties.

DATUM LEVELS.

It may facilitate further surveying operations, should any be undertaken, to place the following information with regard to Datum Levels on record:

The Survey was commenced by different surveying parties at great distances apart, in consequence of which it was impossible to begin the " Levels" with a uniform Datum. Distinct Datums were assumed by each party, and as " Bench Marks" were left in the woods, on each line of survey, with the heights marked thereon for future reference, it was thought best in preparing the Plans and Profiles to adhere to the Datum assumed in each case.

The relative position of each Datum may thus be explained:

First Datum.—On this Datum, levels were carried forward from the Restigouche up the Gounamitz to Green River; here they were taken up and carried forward to the Toledi and Rimouski waters; thence by the Abawisquash to River Trois Pistoles. On this Datum also levels were carried from the Restigouche to the Tobique, then to the Nashwaak and to Keswick Summit.

Second Datum.—On this Datum, levels were carried from a point five miles up Keswick valley to Keswick Summit; also from the same point past Fredericton to Little River.

Third Datum.—On this Datum, levels were carried from Little River to Coal Creek.

Fourth Datum.—On this Datum, levels were carried from Apohaqui Station, on the St John and Shediac Railway, northerly to Coal Creek.

On the close of the Survey these levels were found to be relatively as follows :

High water, River St Lawrence at Trois-Pistoles.................................	70.00 feet.
First Datum, said to be high water at Chatham, on the Miramichi,............	84.81 "
Second Datum..	101.81 "
Third Datum..	58.00 "
Fourth Datum, said to be 100 feet under high water on Bay of Fundy at St. John City..	0.00 "

Any discrepancy which exists in the above levels may be due to various circumstances, partly perhaps to the accumulation of small errors. There is nothing however which can possibly affect the general results of the Survey.

The Datum for the Nova Scotia survey is low water at Parsboro, on the Basin of Minas.

The Datum for the Matapedia survey is high water above Campbelton, or Bay Chaleurs, and on the River St. Lawrence at St. Flavia.

FITNESS FOR SETTLEMENT

AND AGRICULTURAL CAPABILITIES OF THE COUNTRY.

A person who has been accustomed to the fine open hardwood forests of Upper Canada, would at first be unfavorably impressed with the quality of the land in the maritime provinces generally, as well as that portion of Canada east of Quebec, if he judged solely from the appearance of the growing timber. Spruce, of several varieties, grows almost universally, intermixed with other kinds of timber; it frequently attains considerable dimensions, and next to the white Pine, is considered of the greatest commercial value. Immense quantities of Spruce deals are annually exported from New Brunswick.

Black and yellow Birch, woods little known in Canada, but largely used in, and exported from the Lower Provinces, to a large extent take the place of Maple and other hard woods. When birch grows with the spruce and other forest trees, the soil is generally considered of good quality. In some sections of the country a proportion of maple is sometimes found, with birch, spruce and other varieties of timber.

The occurrence of spruce with balsam, so common in the forests of Lower Canada and New Brunswick, presents serious obstructions to exploring and surveying operations, as a view of any part of the country beyond a few yards from the position of the observer, is only obtained with great difficulty.

Perhaps the least favorable portion of the country for settlement, along the general route of the surveyed central line, lies between the waters of the St. Lawrence, and the Restigouche. I have traversed this district in various directions, and although I must confess that its agricultural capabilities do not impress me favorably, yet Mr. Walter Lawson, who spent six consecutive months in charge of one of the surveying parties in this locality, and who is well qualified to judge, thus reports :

"In answer to your questions, as to the quality of the country I have been exploring during the last summer, I beg to state that when we left Rimouski at the end of last May, the spring was fairly commenced, and we found no snow in the woods. That on reaching the boundary line between Canada and New Brunswick, we found vegetation as far advanced as anywhere between that and the St. Lawrence.

"The country we passed through was hilly, with rock cropping out on the sides in a few places, but no bare hills, the highest ground being generally rolling, and well timbered with large Birch, Spruce and Balsam.

"I have explored in Canada from Rimouski Village to the Boundary Line, Store Camp No. 1, at Monument No. 47, near the head waters of the Rimouski River; thence, eastward, seven miles, and round, southerly, to the Forks of Green River in New Brunswick; thence, northerly, along Green River and th head waters of the Toledi to Monument No. 89; also, I have traversed in several directions, the country bounded by Sandy Lake, Eagle Lake and Island Lake on the west, and Abawisquash on the north, the Rimouski on the east, and the twelve mile stretch of the boundary line, from Monument No. 39 to No. 47 on the south. This country generally has been lumbered over, consequently very little pine or heavy spruce was met with. The whole is well watered, and most of it eligible for settlement; in no part did I meet with bad land, and in many places I consider the soil of a superior quality.

"The lower section of the valley of the Abawisquash, near the River Trois-Pistoles, is partly settled, and the lands I have been exploring are fully equal, if not superior, to the best land I saw in that settlement."

The district above referred to, embraces an area of probably 400,000 acres; and the whole of the country south of it to the River Restigouche, as far as my knowledge goes, is similarly timbered.

From the River Restigouche southerly to the Tobique, and from the River St. John easterly to the Sissou Branch, about 40 miles in length by about 30 miles in breadth, the country is generally fit for settlement. In many sections it has a fine intermixture of hard wood timber—and viewing it as a whole, generally it may be considered good second class land, in some places it may be called first rate. I never saw better crops than those which were growing in the settlements on the outskirts of this district. For several miles along the banks of the River Tobique, beds of gypsum crop out, of immense thickness and of excellent quality; it is already drawn away in large quantities and extensively used in the settlements in the State of Maine.

On the lines of survey and exploration between the Rivers Tobique and Miramichi, a growth of Birch, Beech and Maple, with other descriptions of timber, indicate a soil suitable for agricultural purposes. These lines of exploration were about twenty miles apart, and as the intervening and adjoining ground would appear to be in every respect similar, there is no doubt that a great deal of this extensive area is fit for settlement.

From the River Miramichi, on the line surveyed, to the River St. John at Fredericton, there is for the most part a fine growth of hardwood timber, and judging from the portion already cleared along the lower part of the Keswick valley, the soil must be of a superior quality. For a distance of 25 miles northward of Fredericton, the country is already cleared and cultivated.

Between the line surveyed from Fredericton, to the head of Grand Lake and the St. John River, the land is low but of excellent quality. From the Grand Lake, southerly, and over the coal fields, the soil is rather indifferent. Before reaching Apohaqui the line passes through the valley of Studholme Mill Stream; here the soil is very good, producing annually excellent crops of Potatoes, Oats, Buckwheat and Hay.

It is said there is still a great deal of land fit for settlement, and yet unoccupied, between the Grand Lake and the Gulph shore, but its extent I have no means of knowing at the present time. Between Fredericton and the River Restigouche, the land referred to above, adjoining the lines of explorations of last year, and considered generally suitable for settlement, embraces an area of, possibly not much less than 2,000,000 acres. Comparing this extensive tract of land with the soil of Upper Canada, I am inclined to think that it is generally better than any of the unsettled districts in that part of the country.

With regard to the agricultural capabilities of the other sections of New Brunswick, I find a great deal of valuable information on the subject in a report by Professor Johnson, the celebrated Chemist and Agriculturist, made to the Governor of that Province in 1850. The information is so important, and the authority so good, that I have given copious extracts from three out of eighteen chapters, in an appendix hereto.* These extracts refer to the agricltural capabilities of the Province, as indicated by its Geological structure, by a practical survey and examination of its soils, and by the actual yield where settlements are formed.

*See Appendix A.

There remains only to be described the character of the land, and its fitness for settlement in that part of Canada, between the St. Lawrence and the Restigouche along the line of the Matapedia survey.

I find that this subject was specially inquired into some years ago, and a report submitted to the Honorable the Commissioner of Crown Lands of Canada, by Mr. A. W. Sims, the gentleman to whom the enquiry was intrusted. The report embraces all the information desired, and indeed much more than I could give from my own knowledge of the country. I have, therefore, made some extracts and appended them hereto.*

VARIOUS PROJECTED ROUTES.
NEW BRUNSWICK AND CANADA DIVISION OF THE SURVEY.

Having described the Engineering features of the lines recently surveyed and submitted estimates of the quantity of work considered necessary to complete the bridging and grading on each, I shall now refer to all the projected routes which seem worthy of attention, and which possibly may be found practicable on thorough surveys being made.

I do not desire it to be understood that I now report all the lines about to be described as practicable. Some of them I believe to be practicable, but my personal knowledge of others is not sufficient to warrant me in expressing a positive opinion as to their feasibility. The lines and combinations of lines about to be referred to, are those which, from partial examinations and information acquired, I think, offer a reasonable chance of being found practicable ; and they are here described and classified in order that judgment may be formed as to which route or routes may be most eligible for farther surveys.

These lines may conveniently be divided into three classes.

First.—*Frontier Routes.*—Comprising those projected lines which, at one or more points, touch or pass close to the frontier of the United States.

Second.—*Central Routes.*—Those lines which are projected to run through the interior and keep at some distance from the Frontier as well as from the Gulf shore.

Third.—*Bay Chaleurs Routes.*—Comprising those lines which touch the waters of the Gulf of St. Lawrence on the Bay Chaleurs.

The several lines herein referred to may be traced on the accompanying General Map ; they are numbered consecutively from the west to the east. It may be explained that the length of each is ascertained by measuring the distance on the map and adding a certain percentage for curvature. This percentage is based on the difference between the actual chaining of the surveyed lines on the ground, and the length thereof measured on the map. A method of computing the distances, which, although perhaps not strictly correct, appears, under the circumstances, the most accurate that can be adopted ; and it will probably give a sufficiently close approximation.

FRONTIER ROUTES.

Line No. 1.—This line was projected some years ago to connect the Grand Trunk Railway at River du Loup, with the Railway now in operation, from near Woodstock to St. Andrews ; an examination of the country was made by Mr. T. S. Rubidge, in 1859 or '60, and his report, with which I have been favored, contains a great deal of valuable information, much of which is applicable to all the Frontier routes (see appendix C). This line, after leaving River du-Loup, is proposed to follow the valley of River Verte, to the water shed between the St. Lawrence and the River St. John, at an elevation of 880 feet above the sea ; thence in a direction generally parallel to the Temiscouata Road to the falls of the Cabaneau River ; from thence to the head waters of River aux Perches, and by the valley of that stream to the Degelé settlement, at the southerly extremity of Lake Temiscouata. From Degelé the line is proposed to follow the River Madawaska to the River St. John at the village of Little Falls.

From Little Falls this line continues along the easterly bank of the River St. John, which it crosses at Grand Falls. and thence keeps on the westerly bank to Woodstock, connecting with St. Andrew's Railway at some convenient point, probably by way of the Eel River Valley. This line has not been surveyed instrumentally, but it is thought to be practicable ; the only doubtful section is that between River du Loup and the Degelé ; but

*See Appendix B.

should a direct line not be had here, a detour either to the west by the valley of the River St. Francis, or to the east by the River Trois Pistoles, the Ashberish waters and Lake Temiscouata, will, without doubt, be found quite practicable, although the length of the line will be considerably increased thereby.

The estimated distances from River du Loup by this line are as follows:

	RAILWAY.		
	Constructed.	Not constructed.	Total.
To St. Andrews—			
From River du Loup to junction with St. Andrew's Railway.................	....	210	210
Along St. Andrew's Railway.................	67	....	67
Total.................	67	210	277
To St. John—			
From River du Loup to junction with St. Andrew's Railway.................	....	210	210
Along St. Andrew's Railway.................	27	....	27
Surveyed line from St. Andrews Railway by Oromocto to St. John.................	....	82	82
Total.................	27	292	319
To Halifax—			
From River du Loup to St. John as above..............	29	292	319
Railway from St. John to Moncton..............	90	....	90
Moncton to Truro..............	6	109	115
Truro to Halifax..............	61	....	61
Total.................	184	401	585

Line No. 2.—This line is laid down on the Map from River du Loup to a point on the Trois Pistoles River, above the conflucnce of the Abawisquash, where a bridge of an expensive character will be necessary.

The section between River du Loup at this point is common to all the lines about to be described. From Trois Pistoles the line passes over to Lake Temiscouata, by the Ashberish Lake and River; following the westerly shore of Lake Temiscouata to the Degelé settlement, it thence continues along the valley of the River Madawaska to Little Falls and the River St. John, to St. Basil. From this point, instead of following the immediate banks of the St. John to Woodstock, as line No. 1 does, it joins on to the exploration line made some three years ago by the St. Andrews Railway Company, when they seemed to have seriously entertained the idea of extending to Canada. This line leaves the St. John River, near St Bazil, and crosses the Grand River about 10 miles from its outlet; it passes about five miles to the east of Grand Falls and crosses the Tobique about fifteen miles from its mouth; thence it is shown on the Map to cross over by the Otelloch and Munquart Rivers to the St. John at Hardwood Creek. At Hardwood Creek the line crosses the St. John on a bridge proposed to be 100 feet high and fully 800 feet long, and continuing onwards it connects with the existing railway to St. Andrews, at its present Terminus, four miles west of Wodstock.

The reports on the exploration of this line northerly to Little Falls were furnished by the gentlemen representing the New Brunswick and Canada Railway Company, to whom I am much indebted. The detailed information thus obtained will be found on reference to Appendix D. About twenty-seven miles of this line north of Woodstock has been instrumentally surveyed; the remainder to Trois Pistoles has only been partially explored. It is anticipated that serious, although perhaps not insuperable, difficulties will be met with between the high-level crossing of the St. John and the crossing of the River Tobique, as well as near the Degelé on Lake Temiscouata The estimate of the cost per mile, given by the Engineer of the St. Andrews Railway Company, in his report appended hereto, is, I presume, for the grading only.

The estimated distances by this line are as follows:

	Constructed.	RAILWAY. Not constructed.	Total.
To St. Andrews—			
From River du Loup to junction with the present terminus of Canada and New Brunswick Railway, west of Woodstock................	......	223	223
Along Railway to St. Andrews...............	87	......	87
Total........................	87	223	310
To St. John—			
From River du Loup to near Woodstock as above....	......	223	223
Along Railway to proposed western extension from St. John...	45	......	45
Along surveyed line by Douglas Valley to St. John...	......	82	82
Total........................	45	305	350
To Halifax—			
From River du Loup to St. John, as above	45	305	350
Railway from St. John to Moncton...................	90	..	90
" " Moncton to Truro	6	109	114
" " Truro to Halifax................	61	......	61
Total................................	202	414	616

Line No. 3.—From River du Loup to Grand Falls, this line is precisely the same as No. 1. From Grand Falls it crosses over to Salmon River, and there joins the proposed extension of the Canada and New Brunswick Railway, as explored by Mr. Buck, the engineer of that company—(See Appendix D); it then follows Mr. Buck's exploratory line across the Tobique River to the head waters of the Munquart River, thence it crosses over and joins the line surveyed by Mr. Burpee for the New Brunswick Government, from Fredericton to the City of St. John.

This is the most direct line between River du Loup and the City of St. John which is likely to be found practicable. It crosses and recrosses the 'air line,' drawn from the extreme points to the north-easterly angle of Maine, no less than twelve times and does not diverge from it, at any point, more than ten miles. There is, it must be confessed, some little uncertainty with regard to the feasibility of this line, between the forks of the Miramichi and the River Tobique—as well as between the Degelé and River du Loup, these sections having been imperfectly explored; but there is good reason to expect that a careful survey would result in showing that a line not unfavorable might be had through these sections as well as elsewhere. This line would require a very costly bridge over the River St. John near Fredericton, and another over the same river at the City of St. John.

The distances to St. John and Halifax are estimated as follows:

	Constructed.	RAILWAY. Not constructed.	Total.
To St. John—			
From River du Loup to Fredericton	...	235	235
From Fredricton to St. John by Oromocto and Douglas valley	...	66	66
Total..................................	...	301	301
To Halifax—			
From River du Loup to St. John as above............	...	301	301
" St. John to Moncton.....	90	....	90
" Moncton to Truro................	6	109	115
" Truro to Halifax...... ...,.............	61	....	61
Total................,.........	157	410	567

CENTRAL ROUTES.

Line No. 4.—This line is identical with the line surveyed last summer, from the River du Loup as far as Eagle Lake.

From Eagle Lake it follows Eagle stream to the forks of the River Toledi; thence along the general direction of the Squatook Lakes, and across by the head-waters of the Iroquois River to Green River Lake; thence along the most favorable route that can be had to the most westerly branch of the Restigouche, continuing along which, and probably by Hunter's Brook, it may rejoin the line surveyed last summer near the source of Grand River; thence following the surveyed line by Two Brooks, River Tobique, North Branch of the Miramichi and the Keswick valley, to opposite Fredericton. After crossing the River St. John, at Fredericton, it continues along the line of Mr. Burpee's survey from Fredericton to St. John, by Oromocto and Douglas valleys. The only portion of this line not instrumentally surveyed is that between Eagle Lake and Grand River, a distance of perhaps 80 miles. About half this distance, viz : from the Squatook Lakes to the River Restigouche has only been partially explored, but no insurmountable difficulty is supposed to exist. The survey and examinations have shown the whole of the remainder of the line to be entirely practicable.

It must be admitted, however, that the Bridges required across the River St. John at two points, are formidable works.

The distances by this line are estimated as follows :

	RAILWAY.		
	Constructed.	Not constructed.	Total.
To St. John—			
From River du Loup by Island and Eagle Lakes, the Squatook Lakes, Green River Lake, Restigouche, Hunter's Brook and survey line to Fredericton...	....	260	260
From Fredericton, by Oromocto and Douglas Valley to St. John...	....	66	66
Total..	....	326	326
To Halifax—			
From River du Loup to St. John as above...............	....	326	326
From St. John to Moncton....................................	90	....	90
" Moncton to Truro..	6	109	115
' Truro to Halifax..	61	....	61
Total	157	435	592

Line No. 5.—This follows the line surveyed, and already described. From Fredericton to St. John, it follows the Oromocto and Douglas Valley route like Nos. 3 and 4, and equally with them it requires the bridging of the River St. John at two places. With the exception of the portion between Eagle Lake and the sources of the Green River referred to in the description, this line has been instrumentally examined from end to end, and without doubt is quite practicable. The distances to St. John and Halifax are estimated as under.

	RAILWAY.		
	Constructed.	Not constructed.	Total.
To St. John—			
From River du Loup by Island and Eagle Lake, South East Branch of Toledi, North West Branch of Green River, Moose Valley, Gounamitz Valley, Boston Brook, Two Brooks, North Branch of Miramichi and Keswick River to Fredericton...	....	262	262
From Fredericton to St. John by Oromocto and Douglas Valley..	....	66	66
Total..	....	328	328

To Halifax—

	Constructed	Not constructed	Total
From River du Loup to Fredericton as above	....	262	262
From Fredericton to St. John	....	66	66
St. John to Moncton	90	....	90
Moncton to Truro	6	109	115
Truro to Halifax	61	....	61
Total	157	437	594

Line No. 6.—This is identical throughout the whole extent with the line surveyed last summer to Apohaqui Station on the railway running from St. John to Shediac, and need not again be described. The distances by this line are:

RAILWAY.

To St. John—

	Constructed.	Not constructed.	Total.
From River du Loup by Fredericton and head of Grand Lake to Apohaqui	....	343	343
From Apohaqui by Railway in operation to St. John.	37	....	37
Total	37	343	380

To Halifax—

	Constructed.	Not constructed.	Total.
From River du Loup by Fredericton and head of Grand Lake to Apohaqui	....	343	343
From Apohaqui to Moncton	53	....	53
" Moncton to Truro	6	109	115
" Truro to Halifax	61	....	61
Total	120	452	572

Line No. 7.—This line follows precisely the same route as line No 6 from River du Loup as far as the head of Grand Lake, touching on its way the River St. John opposite Fredericton.

From the head of Grand Lake, instead of running southerly to Apohaqui, it continues in a direction nearly easterly, over ground known to be favorable, until it intersects the existing Railway from St. John to Shediac at such point as may be found most suitable, probably about 12 or 13 miles west of Moncton.

The following are the estimated distances to St. John and Halifax by this line:

RAILWAY.

To St. John—

	Constructed.	Not constructed.	Total.
From River du Loup by surveyed line to head of Grand Lake	....	304	304
From head of Grand Lake to Junction with Railway, west of Moncton	....	45	45
Along Railway to St. John	77	....	77
Total	77	349	426

To Halifax—

	Constructed.	Not constructed.	Total.
From River du Loup to head of Grand Lake as above.	....	304	304
From head of Grand Lake to Railway Junction west of Moncton	....	45	45
Along Railway to Moncton	13	....	13
From Moncton to Truro	6	109	115
From Truro to Halifax	61	....	61
Total	80	458	538

Line No. 8.—This line, from River du Loup to the River Restigouche, coincides with the surveyed line (No. 6) between these points. From the Restigouche it follows Boston Brook to Jardines Brook, and continues by an explored passage from the latter stream to the valley of the Sisson Branch of the River Tobique : following which it is supposed that, with some difficulty, a practicable route may be had by the Forks and right hand Branch of the Tobique to Long Lake or Tobique Lake; thence the line is drawn on the map to the sources of Clear Water Brook, and, by a route explored under my directions, by Mr. Tremaine, C. E., in March, last year, to Rocky Brook, and thence by the main Miramichi to Boiestown ; from Boiestown this line is laid down to the head of Navigation on Grand Lake, where it intersects the surveyed line (No. 6) and follows it to Apohaqui Station.

A long extent of this line, viz: from the River Restigouche to Grand Lake, has not been instrumentally surveyed, and it has only been partially explored ; enough, however, is known of the country to give good ground for the supposition that a line may be found, within the limits of practicability, along the general direction of the route indicated.

It is not, however, believed that a line can be had without severe gradients as well as heavy works of construction. Mr. Tremaine's report on the aneroid exploration made by him, from Boiestown across the Tobique Highlands, will be found in the Appendix (E).

The distances to St. John and Halifax by this line are approximately estimated as follows :

	Constructed.	RAILWAY. Not constructed.	Total.
To St. John—			
From River du Loup, by Survey line, to Restigouche.	....	120	120
From Restigouche, by Forks of Tobique and Boiestown, to head of Grand Lake	....	148	148
From head of Grand Lake, by surveyed line, to Apohaqui	....	39	39
From Apohaqui to St. John	37	....	37
Total	37	307	344
To Halifax—			
From River du Loup to head of Grand Lake as above	....	268	268
From Grand Lake to Apohaqui	....	39	39
From Apohaqui, along Railway, to Moncton	53	....	53
From Moncton to Truro	6	109	115
From Truro to Halifax	61	....	61
Total	120	416	536

Line No. 9.—This line follows the same as the last (No. 8), from River du Loup to the head of Grand Lake. From Grand Lake, instead of running to Apohaqui on the surveyed line, it is drawn easterly across a country without engineering difficulties, to a point of intersection with the existing Railway, about 13 miles west of Moncton.

The distances by this line are estimated to be :

	Constructed	RAILWAY. Not constructed.	Total.
To St. John—			
From River du Loup to head of Grand Lake, the same as by line No. 8	....	268	268
From Grand Lake to Railway Junction near Moncton	....	45	45
Along Railway to St. John	77	....	77
Total	77	313	390

	Constructed.	RAILWAY. Not constructed.	Total.
To Halifax—			
From River du Loup to point of intersection west of Moncton with Railway............	....	313	313
Along Railway to Moncton............	13	....	13
From Moncton to Truro............	6	109	115
From Truro to Halifax............	61	....	61
Total............	80	422	502

Line No. 10.—This line corresponds with the two last, Nos. 8 and 9, from River du Loup to the Tobique lakes; it is then drawn across to the village of Indiantown, on a course between the north branch of the Renous River and the Little south-west Miramichi. This route, from the Tobique Lakes to Indiantown, is strongly recommended as favorable, by the Hon. P. Mitchell, of New Brunswick. From Indiantown it follows Major Robinson's line, to Buctouche River, and then continues southerly to Moncton. This is unquestionably one of the most direct lines between Halifax and River du Loup, and possibly it may be found practicable throughout; but it is impossible to speak with certainty, without more information than is at present possessed.

Between the Tobique Lakes, the sources of the Renous and the Miramichi, is the part of the country least known. Mr. Mitchell says that the waters of the Tobique, here interlock with the sources of the Little South-west Miramichi, and that the character of the country is level. This being the case, there is reason to suppose that a railway line may be located through the country on the line indicated.

The distances by this line are estimated as follows :

	Constructed.	RAILWAY. Not constructed.	Total
To St. John—			
From River du Loup to the Tobique Lakes	....	180	180
From the Tobique Lakes to Indiantown............	....	64	64
" Indiantown to E. & N. A. Railway............	....	82	82
" Along Railway to St. John............	96	....	96
Total............	96	326	422
To Halifax—			
From River du Loup to E. & N. A. Railway, as above............	....	326	326
From E. & N. A. Railway to Truro............	....	109	109
" Truro to Halifax............	61	....	61
Total............	61	435	496

Line No. 11.—This line corresponds with the surveyed line (No. 6), from River du Loup to Island lake, and perhaps as far as Eagle lake; It passes over from these waters on a level to the Toradi, and continues along that river up the Rimouski to the boundary line between Canada and New Brunswick; it passes over through a favorable opening in the Highlands to the valley of the south branch of the Kedgwick, and thence it is assumed that the line may gradually descend by the south branch and main Kedgwick to the Restigouche. Difficulties are said to exist in the lower part of the south branch; should these prove too expensive to overcome, they can, I have reason to believe, be entirely avoided by following the general direction of the line shown on the map, from the Restigouche to Kedgwick Lake, and thence down the main valley. From the Restigouche the line is drawn by Five Fingered Brook across to the Sisson branch of the Tobique; here it joins line No. 8, with which it corresponds thence to Apohaqui. On this line difficulties may be encountered in passing over from Five Fingered Brook to the Sisson branch, as well as at points on line No. 8 already mentioned, but it is not supposed they will prove insuperable.

The following are the estimated distances to St. John and Halifax by this line:

	RAILWAY Constructed.	RAILWAY Not constructed.	Total.
To St. John—			
From River du Loup by the Rimouski and Kedgwick, the Forks of Tobique and Boiestown to the head of Grand Lake	....	284	284
From the Head of Grand Lake to Apohaqui	....	39	39
Apohaqui by Railway to St. John	37	....	37
Total	37	323	552
To Halifax—			
From River du Loup to the Apohaqui as above	....	323	323
Apohaqui, along Railway, to Junction	59	....	59
From Junction to Truro	....	109	109
" Truro to Halifax	61	....	61
Total	120	432	552

Line No. 12.—This line is the same as the last, from River du Loup as far as the head of Grand Lake, but here it turns off to the east and intersects the existing Railway a few miles west of Moncton. The distances are estimated to be:

	RAILWAY Constructed.	RAILWAY Not constructed.	Total.
To St. John—			
From River du Loup to the head of Grand Lake, the same as No. 11	....	284	284
From the head of Grand Lake to junction west of Moncton	....	45	45
From junction, along Railway to St. John	77	....	77
Total	77	329	406
To Halifax—			
From River Du Loup to the intersection with the Railway west of Moncton, as above	....	329	329
Along Railway to Moncton	13	....	13
Moncton to Truro	6	109	115
Truro to Halifax	61	....	61
Total	80	438	518

BAY CHALEURS ROUTES.

There lies, south of the River Restigouche, north of the Miramichi, east of the most easterly central line above described, a tract of country over sixty miles in width, and extremely unfavorable for Railway construction. Owing to the rugged and mountainous character of this district, it is hopeless to look for a line suitable for a Railway through it, and in consequence of these features, the lines already referred to, all pass to the west, while those about to be described are led round the other side of this Highland region, as far to the east as the shores of the Bay Chaleurs; hence the name by which the latter lines are designated, to distinguish them from the Central and Frontier Routes.

Line No. 13.—This line continues on the same course as the line, No. 11, from River du Loup, by Island Lake, River Toledi and Rimouski, to Kedgwick Lake. From Kedgwick Lake it is thought the line can be carried into the valley of the Patapedia and thence to the Restigouche. It must be confessed that this is only a conjecture, based not on a knowledge of the immediate locality, as the explorations did not extend to this quarter, but on a knowledge of the general character of the country. Should, however, this view prove incorrect, it is probable that a line may be had a little further north, as shown on the map, to the valley of the Matapedia and thence to the Restigouche.*

Both routes measure about the same length, to a common point on the Restigouche River, at the mouth of the Matapedia. With regard to their respective merits or demerits, a safe opinion cannot be formed without a survey.

At present, all that can be said is, that a favorable communication by one or other of these routes is not improbable. From the mouth of the Matapedia the line follows the route recommended by Major Robinson, to Indiantown on the Miramichi River. From Indiantown it continues nearly due south to the head of Grand Lake, and thence by the surveyed line to Apohaqui.

No serious difficulty is anticipated between Indiantown and Grand Lake.

The distances by this line, from River du Loup to St. John and Halifax, are estimated to be as follows:

		RAILWAY.	
	Constructed.	Not constructed.	Total.
To St. John—			
From River du Loup by Patapedia and Restigouche to Dalhousie..	....	183	183
From Dalhousie to Bathurst..............................	....	53	53
" Bathurst to Indiantown	....	59	59
" Indiantown by head of Grand Lake to Apohaqui.	....	92	92
" Apohaqui along Railway to St. John............	37	....	37
Total...............................	37	387	424
To Halifax—			
From River du Loup by Dalhousie, Bathurst, and Grand Lake to Apohaqui, as above...............	....	387	387
From Apohaqui along Railway to Moncton............	53	....	53
" Moncton to Truro..................................	6	109	115
" Truro by railway to Halifax.....................	61	....	61
Total...............................	120	496	616

Line No. 14.—This line coincides with No. 13 from River du Loup to Indiantown, but from Indiantown instead of running southerly to Apohaqui, it follows a south-easterl course along Major Robinson's line nearly the whole distance to Moncton. The distances by this line are estimated to be:

		RAILWAY.	
	Constructed.	Not constructed.	Total.
To St. John—			
From River du Loup, by Rimouski, Patapedia and Restigouche Rivers, Dalhousie and Bathurst, to Indiantown, the same as by line No. 13..........	....	295	295
From Indiantown to E. & N. A. Railway.............	....	82	82
Along Railway to St. John.............................	96	....	96
Total.................................	96	377	473
To Halifax—			
From River du Loup to E. & N. A. Railway, as above...	....	377	377
From E. & N. A. Railway to Truro.....................	....	109	109
" Truro to Halifax	61	....	61
Total........	61	486	547

* "A party was sent to explore for a line from the Matapedia River, westward, following the valley of one of its tributaries, and thence across to the Rimouski River, and from the reports I have received from them, it appears probable that a practicable line may be obtained by following the valley of Metallics Brook 5 miles below the forks of the Matapedia and along a succession of Lakes to the Rimouski and by the valley of the Torcadia to the Abersquash."—*Captain Henderson's Report.*

Line No. 15.—This is the route known as Major Robinson's line. It runs from River du Loup to the Trois-Pistoles crossing, already referred to, and continues from thence at a distance of eight to 12 miles from the south shore of the St. Lawrence to the River Métis. From the Métis the line passes over to the valley of the Matapedia, which it descends to the River Restigouche. The Restigouche leads it to Bay Chaleurs, the shores of which it follows to the Town of Bathurst, passing on the way the villages of Campbelltown and Dalhousie. From Bathurst the line runs by the Rivers Nepisiguit and the North-west Miramichi to Indiantown on the main or South-west Miramichi. From Indiantown it strikes across a country reported to be flat and favorable, to the Isthmus between the bend of Petitcodiac and Shediac, and thence to Nova Scotia by a route already described.

The recent survey has proved that the Matapedia section will be much less difficult and expensive than was previously supposed.

Instead of twelve or fourteen bridges across the main river, averaging from 360 to 450 feet long, on the first 38 miles north of the Restigouche, only one bridge of 150 feet span is required. Besides which, excavation and other work will be very materially reduced, by adopting curves and gradients, equally as favorable as on other lines of railway both in Europe and America.

The unlooked-for difficulties in the neighborhood of the Metis River have already been referred to ; between this point and Trois Pistoles the country seems to have only been partially surveyed in 1848, as there are other points at which very thorough explorations will require to be made before a location survey can be attempted. The bridging of the Trois Pistoles, common to all lines except No. 1, is a very formidable affair ; that of the Rimouski, where the line crosses at the mouth of the " Ruisseau Bois Brûlé," seemed to me to be not much less so. I think the latter can be avoided, or at least very materially diminished, by a route a little further to the south.

Between the mouth of the Matapedia and Moncton this line will be generally on favorable ground ; and with the exception of the bridges over some of the large rivers, the work, it is expected, will not be heavy.

The distances to St. John and Halifax by this line are estimated to be as follows :

	RAILWAY.		
	Constructed.	Not constructed.	Total.
To St. John—			
From River du Loup, by Metis and Matapedia, to Dalhousie....................................	....	196	196
From Dalhousie to Bathurst................................	....	53	53
" Bathurst to E. & N. A. Railway................	....	141	141
Along E. & N. A. Railway to St. John............	96	....	96
Total..	96	390	486
To Halifax—			
From River du Loup by Metis, Matapedia, Dalhousie and Bathurst to Moncton........................	....	390	390
From Moncton to Truro................................	....	109	109
" Truro by Railway to Halifax................	61	....	61
Total..	61	499	560

The distances by the various routes may now be presented in a Tabular form, and it may be mentioned that the distances here submitted considerably exceed those given by Major Robinson and others ; the allowances which I have made in every case for curvature, and which I deem absolutely necessary in order to insure a safe estimate, may account for this excess. Major Robinson estimated the distance from Halifax to Quebec at 635 miles. By adding the length of the Quebec and River du Loup Railway to the figures now given, the distance by the same route would appear to be fifty miles longer—equal to about eight per cent. on the whole. Should the allowance for curvature (which I am convinced is ample) ultimately prove greater than necessary, the estimates will at least possess the merit of erring on the safe side, and any possible error of this kind will not affect a comparison of the different routes, as, in this respect, all are relatively treated alike.

Table of Comparative Distances from River du Loup to St. John and Halifax.

ROUTES.	No. o line.	TO ST. JOHN.			TO HALIFAX.		
		Railway Built.	Not Built.	Total.	Railway Built.	Not Built.	Total.
Frontier Routes.	1	27	292	319	184	401	585
	2	45	305	350	202	414	567
	3	00	301	301	157	410	561
Central Routes.	4	00	326	326	157	435	592
	5	00	328	328	157	437	594
	6	37	343	380	120	452	572
	7	77	349	426	80	458	538
	8	37	307	344	120	416	536
	9	77	313	390	80	422	502
	10	96	326	422	61	435	496
	11	37	323	360	120	432	552
	12	77	329	406	80	438	518
Bay Chaleurs Routes.	13	37	387	424	120	496	616
	14	96	377	473	61	486	547
	15	96	390	486	61	499	560

With regard to the *Total distance from River du Loup to St. John, including the length of Railway already constructed,* the several lines stand in the following order, beginning with the shortest :

FROM RIVER DU LOUP TO ST. JOHN.

Line No. 3, Frontier Route, ... Total length .... 301 Miles.
" 1, ... " .... " .... 319 "
" 4, Central Route, ... " .... 326 "
" 5, ... " .... " .... 328 "
" 2, Frontier Route, ... " .... 350 "
" 11, Central Route, ... " .... 360 "
" 6, ... " .... " .... 380 "
" 9, ... " .... " .... 390 "
" 12, ... " .... " ...... 406 "
" 10, ... " .... " .... 422 "
" 13, Bay Chaleurs Route,....... " .... 424 "
" 7, Central Route,............. " .... 426 "
" 14, Bay Chaleurs Route,....... " .... 473 "
" 15, ... " .... " .... " .... 486 "

In respect to the length of Railway *yet to be constructed,* to connect *River du Loup with St. John,* the several lines may be placed in the following order :

Line No. 1, Frontier Route, to be constructed ...... 292 Miles
" 3, ... " .... " .... 301 "
" 2, ... " .... " .... 305 "
" 8, Central Route, ... " ... . 307 "
" 9, ... " .... " .... 313 "
" 11, ... " .... " .... 323 "
" 4, ... " .... " .... 326 "
" 10, ... " .... " .... 326 "
" 5, ... " .... " .... 328 "
" 12, ... " .... " .... 329 "
" 6, ... " .... " .... 343 "
" 7, ... " .... " .... 349 "
" 14, Bay Chaleurs Route,....... " .... 377 "
" 13, ... " .... " .... 387 "
" 15, ... " .... " .... 390 "

Comparing the distances from River du Loup to Halifax and including the length of Railway already constructed, the table shows that the several lines stand in the following order:

	Line No.	Route	Total length	Miles
	10,	Central Route,	Total length	496 Miles.
"	9,	"	"	502 "
"	12,	"	"	518 "
"	8,	"	"	536 "
"	7,	"	"	538 "
"	14,	Bay Chaleurs Route,	"	547 "
"	11,	Central Route,	"	552 "
"	15,	"	"	560 "
"	3,	Frontier Route,	"	567 "
"	6,	Central Route,	"	572 "
"	1,	Frontier Route,	"	585 "
"	4,	Central Route,	"	592 "
"	5,	"	"	594 "
"	2,	Frontier Route,	"	616 "
"	13,	Bay Chaleurs Route,	"	616 "

Comparing again the distance to Halifax, having in view simply *the length of Railway yet to be built*, the several lines would stand as follows :

	Line No.	Route		Miles
	1,	Frontier Route, to be constructed		401 Miles.
"	3,	"	"	410 "
"	2,	"	"	414 "
"	8,	Central Route,	"	416 "
"	9,	"	"	422 "
"	11,	"	"	432 "
"	10,	"	"	435 "
"	4,	"	"	435 "
"	12,	"	"	438 "
"	6,	"	"	452 "
"	7,	"	"	458 "
"	14,	Bay Chaleurs Route,	"	486 "
"	13,	"	"	496 "
"	15,	"	"	499 "

From the foregoing the following deductions may be drawn :

Line No. 3 is the shortest Frontier Route *to St. John* ; its total length is 301 miles, the whole of which is yet to be built. By this line the total distance to Halifax is 567 miles, of which 157 miles are constructed, leaving 410 miles yet to be made.

Line No. 4 is the shortest Central Route *to St. John* ; its total length is 326 miles, the whole of which has to be made. By this line the distance to Halifax is 592 miles, of which 157 miles are built, leaving 435 miles to be constructed.

Line No. 13 is the shortest Bay Chaleurs Route *to St. John* ; its total length is 424 miles, of which 37 miles are constructed, leaving 387 miles to be made. By this line the total distance to Halifax is 616 miles, of which 120 miles are already made, leaving 496 miles to be built.

Line No. 3 is the shortest Frontier Route *to Halifax* as well as to *St. John*, the distances are already given.

Line No. 10 is the shortest Central Route *to Halifax* ; the total distance by it is 496 miles, of which 61 miles are built, leaving to be built 435 miles.

The total distance to St. John by line No. 10 is 422 miles, of which 96 miles are built, leaving to be constructed 326 miles.

Line No. 14 is the shortest Bay Chaleurs Route *to Halifax* ; its total length is 547 miles, of which 61 miles are constructed, leaving 486 miles to be made. By this line the total distance to St. John is 473 miles, of which 96 miles are built, leaving 377 miles yet to be constructed.

The shortest of all the Lines to St. John is No. 3, Frontier Route.
The shortest of all the Lines to Halifax is No. 10, Central Route.

Line No. 3 requires the construction of 25 miles less than No. 10, to connect River du Loup with both St. John and Halifax; but the total distance *to Halifax* by line No. 3, is 71 miles greater than by line No. 10, whilst the total distance *to St. John* by line No. 10 is 121 miles greater than by line No. 3.

The shortest route from River du Loup to the Atlantic Sea Board, on British Territory is by line No. 1 *to St. Andrews.*

The total distance *to St. Andrews* by this line is estimated at 277 miles, of which 67 miles are constructed, leaving only 210 miles to be built.

The total distance *to St. John* by line No. 1, is 319 miles, of which 292 miles require to be made.

The total distance *to Halifax* by line No. 1 is 585 miles, of which 401 miles require to be built.

DISTANCE FROM THE FRONTIER.

I shall now, in accordance with my instructions proceed to give the distances of the several lines from the Frontier of the United States

Line No. 1 runs immediately along the boundary line, for a distance of about 40 miles; and for a further distance of about 80 miles it ranges from three to twelve miles from the Frontier.

Line No. 2 almost touches the boundary of Maine at two points; one about ten miles northerly from Woodstock, the other between St. Basil and Little Falls. For a distance of 120 miles this line will average not more than eight miles from the boundary.

Line No. 3 runs along the boundary of Maine for about 40 miles, and then gradually diverges from it.

Line No. 4, for a distance of twenty or thirty miles, is within 18 miles of the boundary line.

Lines Nos. 5, 6 and 7 are generally not nearer to the boundary line than the minimum distance between the Grand Trunk Railway and the northern Frontier of Maine; this distance, in a direct line, is from 27 to 28 miles. At one point, lines Nos. 5, 6 and 7 are within this distance, but it is believed that at this point the distance may be increased in making a location survey. Line No. 5 runs from Fredericton to the City of St. John, on the westerly side of the St. John River. Lines Nos. 6 and 7 do not cross the river.

Lines Nos. 8, 9 and 10 are each, only at one point, within 27 miles of the boundary line; throughout the remainder of their course they are at a greater distance from it.

Lines Nos. 11 and 12 are each about 30 miles from the boundary line, at the nearest point, for the rest of the way they are at a much greater distance from it.

Lines Nos. 13, 14 and 15 are each nearer the boundary line at River du Loup than at any other point, and as they run by the Bay Chaleurs, they are generally at an extreme distance from the Frontier of Maine.

COMMERCIAL ADVANTAGES OF DIFFERENT ROUTES.

The next topic upon which I am required under my Instructions to report, is the comparative advantages of the various routes embraced in the survey, in a commercial point of view. In approaching this subject I must confess my entire inability to discuss it satisfactorily. My time has been so wholly taken up with matters purely connected with the survey, during the short period which has elapsed since it commenced, that I have not been able to give this most important question the attention which it justly demands. In my desire, therefore, to carry out the instructions of the Government, I can only submit the imperfect impressions which I have formed on this branch of the enquiry.

It would be needless to attempt a comparison of the commercial merits of each of the fifteen separate lines and combinations of lines herein alluded to; it will probably be sufficient to deal with them generally, as *Frontier, Central* and *Bay Chaleurs* Routes. The Nova Scotia Division of the survey, being common to all routes through New Brunswick, will not be embraced in the comparison; and the military objections to the Frontier lines, or to any of the lines, will, for the present, be disregarded. The question of "Local" and Through traffic" will be considered separately.

LOCAL TRAFFIC.

The valley of the River St. John is generally well settled from the Bay of Fundy to Little Falls, where the Temiscouata Portage to River du Loup (about 75 miles in length) begins.

The lumbering operations of New Brunswick are now carried on, chiefly on the upper waters of the River St. John; and the supplies for the lumbermen, which are not produced in the locality, are now in a great measure brought from the United States, by water to the city of St. John, and thence up the river. A railway from River du Loup, through this section, would enable provisions for consumption in the lumbering districts, not only of New Brunswick but also of Maine, to be brought in direct from Canada, and thus greatly tend to develope the industry and resources of these regions. At the present time, Canadian flour may be seen within sixty miles of the St. Lawrence, after having been transported, in the first place, to New York or Portland, then shipped to St. John and floated up the river in steamers and flat boats. This trade would manifestly be changed by the construction of the Intercolonial Railway, on a frontier route, to the advantage of the lumbering interests; and the traffic resulting therefrom, would form an item in the revenue of the contemplated work. It is said that as much as 80,000 barrels of flour, pork and other merchandize are annually imported to the valley of the River St. John, north of Woodstock; and that the population of this district. including the Aroostook lumbering country in the State of Maine, is estimated at 40,000.

A central route will have the least population to accommodate immediately along the line; indeed between the Miramichi and St. Lawrence there is only one settlement, which consists of a few families on the Tobique River. By opening roads, however, to the east and west, the county of Restigouche and the valley of the St. John would be easily reached, and a considerable portion of the trade of these sections brought within the influence of the railway. A line through the centre of New Brunswick, would take the supplies for the lumbering trade, and would rapidly settle up the large tracts of cultivable land in this district. A railway so situated would, as a line of communication, have similar effects on the trade and progress of New Brunswick as the River St. John has had, with this additional advantage, it would be open all the year, instead of half of it.

In much less time, it is believed, than has been occupied in settling and improving the lands which nature made accessible by the River, would the artificial means of communication result in populating the interior of the country through the greater part of its length; and thus develope and foster a traffic which does not now exist.

A Railway constructed by the Bay Chaleurs would pass through a country already in part settled; and it would be of the greatest importance to Campbellton, Dalhousie, Bathurst, Chatham, and other towns and villages on the Gulf shore. Compared with the Central and Frontier Routes it would not perhaps to the same extent serve the lumbering interests of New Brunswick; nor would it reclaim as much wild land, although there are large sections even on this route said to be capable of cultivation, yet lying wild.

A proper judgment of the local traffic at present existing may, perhaps, best be formed by comparing the population along each route.

The population in the section of country through which a Frontier line would pass, embracing the whole of the counties of Victoria, Carleton, York and one half of Sunbury and Queens, is, according to the last census, 51,175; to which may be added 20,000 for the northern and eastern parts of Maine, which adjoin New Brunswick, and which would be accommodated by a Railway running along its border. If to the above we add the population on the Temiscouata Portage, and a percentage for natural increase since the last census was taken, we shall have a population of over 80,000 in the district which would be served by a Frontier route.

The population in the district affected by the Central routes, is chiefly confined to the section south of the Miramichi, and may be estimated at one half of the counties of Queens, Sunbury and York, amounting to 21,404; to this may be added the whole of the counties of Victoria and Restigouche, 12,575, and a portion of the north-easterly part of Maine; making in all a population of perhaps 40,000, not all directly, but all in some degree accommodated by the construction of a central line.

A line by the Bay Chaleurs would pass through the counties of Kent, Northumberland, Gloucester and Restigouche, in New Brunswick, as well as Bonaventure and Rimouski, in Canada. The population of these six counties amounted to 88,541 when the last census was taken; a limited portion of the county of Gaspé and the natural increase may make the whole population over 90,000.

From the above data, the average number of inhabitants for each mile of Railway by the different routes would be nearly as follows:

   A Frontier   line 260 per mile of Railway
   A Central   " 122   "   "
   A Bay Chaleurs "   235   "   "

With regard to local traffic, therefore, it would appear from the above, that the Railway would receive the largest proportion if constructed on a Frontier Route, and least if constructed on a central route.

Taking population as the basis of computation of local traffic, the average per mile in the country between the River du Loup and the northerly boundary of Nova Scotia, on the completion of the Intercolonial Railway, would compared with that of Canada and the United States, be in the following ratio, nearly:—The whole of New Brunswick and that part of Canada east of the River du Loup, 534 per mile of railway (proposed).
The whole of Canada........................1330   "   " (constructed.)
The whole of United States, about 1000   "   "   "

This may give some idea, although perhaps an imperfect one, of the comparative value of the local traffic which may reasonably be expected on the opening of a line of Railway through the Country.

THROUGH FREIGHT TRAFFIC.

A distinction must necessarily be drawn between "through freight" and "through passenger" traffic; as the former will naturally seek the nearest channel to an open Atlantic port, while passengers for Europe will generally take the route by which they can reach their destination soonest, and that may not be by the line which leads to the nearest Harbour.

The ports of Montreal and Quebec, when open to sea-going vessels, are undoubtedly the most convenient for the shipment of heavy freight from Canada to Europe, but these are periodically closed during the winter season, and are therefore unavailable for over five months in the year.

By the projected lines for the Intercolonial Railway, St Andrews and St. John, on the Bay of Fundy, are the nearest open winter ports to Canada, within British territory, and they would, therefore, be the most available outlets for Canadian produce while other nearer ports remain closed.

At the present time Canadian produce may be shipped during winter, without restrictions, at United States ports; and in the event of the existing treaty arrangement being continued, it would become a question whether United States ports on the Atlantic sea board, or British Ports on the Bay of Fundy, were the easiest reached during the winter months.

The nearest United States port to Toronto is New York, the nearest to Montreal is Portland, and the shortest distances between the several ports referred to are as follows:

  From Toronto to New York direct .... 540 miles.
     "   to St. Andrews by River du Loup............ 889   "
     "   to St. John by River du Loup 913   "
  From Montreal to Portland direct 297   "
     "   to St. Andrews by River du Loup............ 559   "
     "   to St. John by River du Loup 583   "

It is evident, therefore, from the favorable position of New York and Portland, that they will continue to be the most convenient winter outlets for Canadian freight, so long as the Government of the United States abstains placing restrictions on Canadian commerce

In the event, however, of Canadian traffic being prevented from passing through the United States, the Intercolonial Railway wou'd carry, during winter, all the freight to and from the sea board which would bear the cost of transportation; and as the cost would to a great extent, depend on the length of railway to be passed over, it would be of considerable importance to have the shortest and most favorable line, selected, to the best and nearest port on the Bay of Fundy; and therefore, with respect to the "through freight" traffic, the frontier lines are entitled to the preference, and next to them some of the central lines.

As the probable "Through freight traffic" depends on so many contingencies, it is impossible to form any proper estimate of its value : but of this we may rest satisfied, if the construction of the Intercolonial Railway, by opening out an independent outlet to the ocean, prove instrumental in keeping down the barriers to Canadian trade which our neighbours have the power to erect, it might, in this respect alone be considered of the highest commercial advantage to Canada. It is scarcely likely that the people of the United States, would permanently allow themselves to place restrictions on Canadian traffic, when they discovered that by so doing they were simply driving away trade from themselves; and in this view the contemplated Railway may fairly be considered, especially by the people of that part of Canada, west of Montreal, of the greatest value to them when least employed in the transportation of produce to the sea board.

THROUGH PASSENGER TRAFFIC.

The spacious Harbour of Halifax, open at all seasons of the year, is universally admitted to be in every respect suitable for the Terminus of the Intercolonial Railway And here it is supposed that passengers for Europe would embark, in preference to other points from which Ocean Steamers at present take their departure.

Halifax is 550 miles nearer Liverpool than New York; 357 miles nearer than Boston ; 373 nearer than Quebec, and 316 miles nearer than Portland. And doubtless the shortening of the ocean passage by these distances would, to many travellers, be a great object, if proper facilities for reaching Halifax were provided.

The construction of the Intercolonial Railway would enable Canadian Passengers to reach Halifax easily. And on its completion the mail steamers would no doubt run from Halifax in place of Quebec or Portland. New York passengers, on the other hand, could scarcely be tempted to go round by Montreal and River du Loup to Halifax, a distance of nearly 1200 miles, in order to save 550 miles by water. The advantages of a shorter Ocean passage are, however, considered so great by the people of the United States, that a Railway to reach Halifax, by the shortest line, would soon be established ; more especially when the construction of the Intercolonial Railway would connect St. John with Halifax, by way of Moncton and Truro, and leave only the link between St. John and Bangor to be built. Bangor is the extreme easterly extension, as yet, of the American system of Railways. The distance thence to St. John by the route contemplated, and in part surveyed, is estimated at 200 miles. The construction of this link, is most warmly advocated in the State of Maine and in the Province of New Brunswick; already, public aid from both countries has been offered to secure its construction, and the influences and agencies at work will, I am convinced, be instrumental in finishing this line of communication at no distant day—perhaps simultaneously with, or possibly before, the completion of the Intercolonial Railway

It would obviously be unwise, therefore, to overlook this projected route in forming estimates of probable traffic on the Intercolonial line.

The United States Route by Bangor would intersect the Grand Trunk Railway at Danville Station, 28 miles out of Portland. and thus form an unbroken railway connection, having the same width of track from Halifax to Montreal and all other parts of Canada. The distance from Halifax to Montreal by this route is estimated at 846 miles, while the distance by the Frontier and Central lines, which form the shortest connection between Canada and the Bay of Fundy, embracing lines Nos. 1 to 6, averages 871 miles in length. Thus, it is evident that the passenger traffic of the Intercolonial may, on any of these lines being constructed, be tapped near its root, and much of it drawn away.

Under these circumstances, it is too apparent that the Intercolonial Railway may find in the United States route, a formidable rival for Canadian passenger traffic, to and from Europe, by way of Halifax.

Fortunately, with a view to counteract this difficulty, a line by the Bay Chaleurs would offer special advantages, which may here be noticed.

The Chart which accompanies this will show that the entrance to the Bay Chaleurs is so situated, geographically, that while it is about as near Europe as the entrance to Halifax harbour, it is, at the same time, several hundred miles nearer Montreal and all points west of that city.

Some of the projected lines of Railway touch the Bay Chaleurs at Dalhousie and at Bathurst; the latter place is not admitted to be suitable for the purposes of steam navigation, and the former, although in possession of a fine sheet of water, well sheltered and accessible at all conditions of the tide, is, nevertheless, from its position at the extreme westerly end of the Bay, farther inland than might be wished. In order to reduce the steamship passage to a minimum, it is desirable to have the point of embarkation as far easterly as possible, and therefore the existence of a commodious harbour near the entrance of the Bay is of no little importance. A place named Shippigan, on the southerly side of the entrance of the Bay Chaleurs, appears to have many of the requisites of a good Harbour. It is thus spoken of in the reports on the Sea and River Fisheries of New Brunswick,* published under the authority of the Legislature of that Province.

"GREAT SHIPPIGAN HARBOUR.

"This spacious harbour is formed between Shippigan and Pooksoudie Islands and the main land. It comprises three large and commodious harbours: first, the great inlet of Amqué, in Shippigan Island, the depth of water into which is from four to six fathoms; second, the extensive and well-sheltered sheet of water, called St. Simon's Inlet, the channel leading to which, between Pooksoudie Island and the main, is one mile in width, with seven fathoms water from side to side.

"The principal entrance from the Bay Chaleurs has not less than five fathoms on the bar, inside which, within the harbour, there are six and seven fathoms, up to the usual loading place, in front of Messrs. Moore and Harding's steam saw mill at the village; from thence to the gully there is about three fathoms water only. Vessels within the harbour of Shippigan have good anchorage, are quiet safe with every wind, and can load in the strongest gale. The rise and fall of the tide is about seven feet.

"The noble haven called St. Simon's Inlet, the shores of which are almost wholly unsettled and in a wilderness state, runs several miles into the land, maintaining a good depth of water almost to its western extremity."

Duncan McNiel, an old pilot, frequently employed on the Government steamers when calling at New Brunswick ports, describes Shippigan as a good harbour, with plenty of water, regular soundings and tough blue clay-holding ground, indeed where vessels would be perfectly secure in any storm. He says that he could take a ship of heavy draught into it in any weather, by night or by day; that in dirty or dark weather he would go entirely by the lead.

Others describe Shippigan harbour as unobjectionable. The Admiralty chart seems to agree in the main with the descriptions above given; it shows that the area of the basin, embracing only the water over the three fathom line at low tide, is about two and a half square miles; a sheet about double the size of Halifax harbour between St. George Island and the narrows to Bedford Basin. The only objectionable feature seems to be the channel at the entrance, which is about three miles long to the basin, a little crooked, and at present without leading marks; is is, however, about half a mile in width, free from all obstructions, the depth varying from five to nine fathoms at low water. There is good warning by the lead in the channel and the approaches to it.

It would appear from the above, therefore, that Shippigan Sound presents a favorable opportunity for forming a traffic connection between the Intercolonial Railway and Ocean Steamers.

A comparison of distances, will now show the importance of Shippigan, in connection with the contemplated Railway:

DISTANCE TO LIVERPOOL.

	Miles.
From Halifax, by Cape Race............	2466
From Shippigan, by Cape Race............	2193
From Shippigan, by Belleisle............	2318
Difference against Shippigan, by Cape Race............	27
Difference in favour of Shippigan, by Belleisle............	148

*By Mr. H. Perley, late Her Majesty's Emigration, and latterly Fishery Commissioner.

DISTANCE TO QUEBEC.	Miles.
From Halifax, by Bangor and Danville	865
From Halifax, by Bay Chaleurs route	685
From Shippigan, by Bay Chaleurs route	419
Difference against Halifax by Intercolonial line	266
Difference against Halifax by United States line	446

DISTANCE TO MONTREAL.

From Halifax, by Bangor and Danville	846
From Shippigan, by Intercolonial route	575
Difference against United States route	271

DISTANCE TO TORONTO.

From Halifax, by Bangor and Portland, Boston, Albany and Niagara Falls	1300
From Shippigan, by Intercolonial line and Canadian Railways	906
Difference against United States routes	304

DISTANCE TO BUFFALO.

From Halifax, to Bangor, Portland, Boston and Albany	1210
From Shippigan, by Intercolonial and Grand Trunk to Toronto, and by Great Western to Niagara Falls and Buffalo	1012
Difference in favor of Intercolonial and Canadian Routes	198

DISTANCE TO DETROIT.

From Halifax, by Bangor, Portland, Boston, Albany, Buffalo and Cleveland	1572
From Halifax, by Bangor, Portland, Boston, Albany, Niagara Falls and Great Western Railway	1446
From Shippigan, by Intercolonial and Grand Trunk Railways	1137
Difference in favor of Shippigan and against United States Route.	435
Difference against Unites States and Great Western	309

DISTANCE TO CHICAGO.

From Halifax, by Bangor, Portland, Boston, Albany, Buffalo, Cleveland and Toledo	1748
From Shippigan, by Intercolonial line, Montreal, Toronto and Detroit	1418
Difference in favor of Shippigan and against United States Route.	330

DISTANCE TO ALBANY.

From Halifax, by Bangor, Portland and Boston	912
From Shippigan, by Intercolonial, River du Loup and Montreal	817
From Shippigan, by Intercolonial (line No. 13) to Apohaqui, then by St. John, Bangor, Portland and Boston	879
Difference in favor of Shippigan and Intercolonial by River du Loup	95
Difference in favor of Shippigan and Intercolonial Route by Apohaqui	33

DISTANCE TO NEW YORK.

From Halifax, by Bangor, Portland and Boston	943
From Shippigan, by Intercolonial line to River du Loup, thence by Grand Trunk to Sherbrooke and by Connecticut River Railway*	927

* This route will be complete on the construction of a Railway now in progress, and some 30 miles in length, by the Massiwippi Valley. This short Railway will connect the Grand Trunk, south Sherbrooke, with the Connecticut River line and form a direct route to New York.

DISTANCE TO NEW YORK.—*Continued.*

	Miles.
From Shippigan, by Intercolonial (line No. 13) to Apohaqui, thence by St. John, Bangor, Portland and Boston	910
Difference in favor of Shippigan and Intercolonial route by River du Loup	16
Difference in favor of Shippigan and Intercolonial Route by Apohaqui and St. John	33

DISTANCE TO ST. JOHN, N. B.

From Halifax, by Moncton	266
From Shippigan, by Apohaqui	233
Difference in favor of Shippigan	33

The above comparisons show that while Shippigan is practically not farther from Liverpool than Halifax, Halifax is farther from the various places referred to as follows:

	Miles.
From Quebec, by Intercolonial route	266
From Quebec, by United States Route	446
From Montreal, and all parts west on the Grand Trunk, by the Intercolonial	266
From Montreal, by the United States Route	271
From Toronto, " "	394
From Buffalo, " "	198
From Detroit, " "	435
From Detroit, by the United States and the Great Western Railway	309
From Chicago, by the United States	330
From Albany, " "	95
From New York " "	16 and 33

The above distances also show that Shippigan is 33 miles nearer St. John, N. B., Portland, Boston, New York, and every point west, by the Intercolonial line to Apohaqui, than Halifax is by the shortest possible route now contemplated.

It is obvious, therefore, that the adoption of Shippigan as the point of connection with Ocean Steamers, would not only neutralize the danger to be feared from the rivalry of the Bangor extension, but it would constitute this line, as far as it could bring traffic, a feeder to the Intercolonial Railway from the south. It is clear too, that the extremely favorable position of Shippigan, in relation to the whole of New Brunswick and Canada, as well as to all points in the Western States, bordering on, and west of the Great Lakes, would prove most beneficial to the Intercolonial Railway, in securing to it a very large share of "Through Passenger Traffic."

It is true that this port on the Bay Chaleurs could only be used probably during seven or eight months in the year, as the Gulf of St. Lawrence cannot be considered navigable during the winter season. But as the great majority of passengers, including emigrants, travel during the summer, the Intercolonial Railway would be situated in a most favorable position for carrying them. It would also, without doubt, have a reasonable chance of securing the transportion of the great bulk of European Mail matter, as well as all descriptions of light Express freight, which usually seeks a rapid means of transit. During a great part of winter Halifax would be the point of connection between the steamers and the proposed Railway; then the latter would unavoidably enter into competition with the United States lines.

There is this objection to the selection of Shippigan as the port of call for Ocean steamers, it would involve the construction of 45 miles of additional Railway. This is not, however, at present indispensable, as Dalhousie could be advantageously used, until circumstances justified the building of a branch from the main line to Shippigan.

The touching at this port on the Gulf, would probably result eventually, in other special advantages, national as well as commercial, the nature of which are more particularly referred to in the Appendix (F).

In summing up the foregoing, it is obvious that, as far as I am capable of judging, the comparative advantages of the various routes may thus be stated:

A Frontier Route would accommodate the largest amount of "Local" traffic, and in the highest degree would serve the purpose of Canada in winter as an outlet for heavy "THROUGH FREIGHT."

A Central Route would, next to a Frontier line, be the best for the transportation of "THROUGH FREIGHT;" and, as a means of colonizing the Country and developing its natural resources, would stand in the first position.

A Bay Chaleurs Route would best secure the largest European "*Passenger Traffic,*" the carriage of *Mail matter* and *Express Freight*, and, next to a Frontier line, would accommodate the greatest amount of "Local" traffic.

Before it can be decided which of these advantages preponderate, and which route is entitled to the preference, the whole subject ought to be carefully and deliberately weighed in all its bearings. I am not, however, called upon to decide this point, and therefore I refrain from expressing an opinion. Indeed, I may add, that the foregoing observations are submitted, with no little hesitation and reluctance, in consequence of the sectional difficulties, which, without doubt, surround this branch of the subject. I could not, however, avoid reference to the commercial merits of the several routes, without disregarding my instructions; and in endeavoring to comply with the wishes of the Government,* it was impossible for me to overlook the main points, which above are imperfectly presented.

CLIMATIC DIFFICULTIES.

Experience has shown that the climate of British North America has a peculiar effect on the works of construction of Railways, as well as on the degree of facility with which they may be maintained and operated after completion,—And as the remedies which may be applied to guard against and counteract the unfavorable influences of climate, to a considerable extent affect the expenditure on construction, I shall before entering on the consideration of the probable cost of the proposed undertaking, allude briefly to this subject.

The frost in these Provinces is in winter very severe. It penetrates the ground where denuded of snow to a depth of several feet, sometimes it is said, in extreme cases, to as much as three and four feet. On exposed points such as the slopes of cuttings and embankments, the snow is sometimes drifted away by the wind, and on the rail track it has always to be removed by artificial means to allow the passage of trains. At such points where the surface is unprotected by a covering of snow, the frost has a free opportunity to penetrate; and if, owing to the springy and spongy nature of the soil, water is retained in such places, the effects of freezing and thawing are frequently very damaging.

Embankments made of certain kinds of earth whilst fresh and loose, naturally take up and hold a good deal of the rain fall of autumn, which is frozen solid during the ensuing winters; they are in consequence exposed to trials when the thaws of spring set in, and frequently considerable outlay is required to restore them to their original and proper shape. It is desirable, therefore, that these sources of outlay should be anticipated and sufficient provision made for them in the first instance. Unless this be done, disappointment at the excessive cost of maintenance of the works will inevitably arise; and however faithfully the parties engaged in the construction may have endeavored to execute their duties, they will be exposed to reflection of an unsatisfactory nature, whilst the causes for such dissatisfaction, instead of being due to negligence or unskilfulness may be solely due to climatic influences. It is essential therefore that provision should be made for expenses of this character until the earth works attain that solidity and sufficient degree of imperviousness, which time alone can give them.

The first and second winters, with the thaws of the following springs, are the most trying on new embankments, but after the third year there is ordinarily little or no difficulty or expense.

Cuttings through wet springy soils are not so soon rendered firm and stable. Year after year on the breaking up of winter, the fresh thawed soil will frequently be in a semifluid state, and in this condition will flow into the ditches, sometimes across the bottom of the "cut," covering in "slurry" the ballast, ties and rails. This is a yearly occurrence in

*Letter of the Honorable the Provincial Secretary, Quebec, 7th May, 1864.

many of the cuttings on the existing Railways in Nova Scotia, and it is no doubt due to the peculiarity of soil and climate here alluded to.

The road bed itself, even when moderately well ballasted, is often greatly disturbed by the effects of freezing and thawing, and the track is thrown thereby out of its uniform level, producing an irregularity of surface alike damaging to the rails, rail-fastenings and rolling stock. It is impossible, moreover, with the track in this condition to maintain the speed of trains with a due regard to safety. These effects on the road bed and track are not confined to cuttings, for they are sometimes witnessed on level sections of country; but they are invariably attributable to the undue presence of water in the soil, within the frost limit. Ditching to some extent obviates this difficulty, but as usually practised in this country, it is not a complete remedy for these evils; true it has the effect of taking off the water from the surface, but it does not remove that which lies under the surface, and which, when acted upon by frost, is equally damaging. I am satisfied that in this latitude not only must the surface water be removed, but that, for the permanent benefit of the Railway, the sub-soil must be kept dry by a system of thorough under-draining. By such a system it is proposed to remove all springs or standing water as well as all soakage from the surface for a depth which exceeds the extreme frost limit; and thus it is believed an effectual remedy will be provided for this particular climatic difficulty and render the slopes of cuttings and the road bed permanently dry and solid.

In all works of masonry, in contact with the earth, care must necessarily be taken to guard against the expansive power of frost; and in the construction of bridges over rivers subject to heavy freshets and flows of ice, more than ordinary precautions must be taken to insure the stability of the structures.

The climate of this country requires that to operate the line efficiently, the utmost care must be taken to insure an abundant supply of water for the engines, not liable to be frozen up during the winter months; without which it will be impossible to operate the line of Railway satisfactorily. The provision of an efficient frost proof Water-service may therefore be considered indispensable.

But the chief climatic difficulty to contend with on the route of the proposed railway is *snow*; to obviate this difficulty is a question of the utmost importance, as upon it mainly depends the value of the Intercolonial Railway as a winter means of communication. The snow-fall along the route of the Intercolonial Railway, according to information received, is very variable. In Nova Scotia and the southern part of New Brunswick, as a general rule it would appear that the snow does not remain on the ground to a greater depth than it ordinarily does in Upper Canada. Probably, however, the snow-fall, although in the aggregate fully greater than in Upper Canada, is more variable than in that Province. Heavy falls of snow are frequently followed by sudden thaws in Nova Scotia, so that the ground is left in certain districts comparatively bare; at other times and places the snow will remain to a considerable depth.

In the central and northern parts of New Brunswick, and northerly to the St. Lawrence, the snow invariably remains on the ground from the beginning to the end of winter. The average depth in the woods where it is not affected by drifting, will range from three to four feet; occasionally, I am told it will reach as much as five feet, sometimes even a greater depth, but as these latter cases are not so well authenticated, I must treat them as exceptional.

In the winter of 1863-1864, so far as my own observations go, the average depth was a little over three feet. During the present winter I believe it is about 4 feet,—that is to say in the woods. In the settlements the dry snow is constantly exposed to drifts and it frequently accumulates to very great depths; on meeting with obstructions it will be found deposited sometimes to twelve and fifteen feet in depth.

Snow drifts, where they happen to occur, are serious obstacles to Railway operations; they are found to be the cause of frequent interruptions to the regular running of trains, besides often the necessity of a heavy outlay. Every winter in Lower Canada the trains are delayed for days at a time on account of these drifts, the mails are in consequence stopped and traffic is seriously interfered with.

Experience goes to prove that these snow drifts only occur where the country is settled, and the surface denuded of its timber; in such places, what are termed "snow fences" have been erected along the railway lines, but these, besides being only temporary expedients, do not always prevent the line of communication from being blocked up with snow. I am convinced that the only effectual method to prevent snow drifts is to follow the plan which nature herself suggests. *There are no drifts in the woods;* the standing timber prevents the snow from being moved by the wind after it falls. It seems, therefore, only necessary to leave a belt of woodland along the line of railway, where it passes through the forest and to cultivate through cleared districts, a second growth of Spruce or Balsam trees, to a width along the railway route sufficient to arrest the drifting snow on the outer side, at a safe distance beyond the limits of the line of traffic. With such provision I believe there would be nothing to fear from *drifts*, even in this high latitude, and it only remains to be considered how the *even* snow-falls ranging from three to five feet on the level may be dealt with.

Although five feet of snow is, perhaps, an extreme average depth, and not frequently occurring where drifts are not common, I consider it highly important, in order that communication may be kept up with satisfactory regularity at all seasons, to provide, if it be possible, for operating the road even when unusual snow-falls occur.

A depth of five feet of snow would, on railways as they are ordinarily made in this country, render it extremely difficult and expensive to operate them; long and narrow cuttings would become so completely blocked up that they could only be opened by a slow process of manual labor, and frequent delays and serious interruptions would be the consequences.

The true way to meet these difficulties, in my opinion, is to adopt a form of construction which will afford the readiest opportunity for the removal of the snow as it falls, by the help of steam power. A fall of snow on *an embankment* is easily removed; snow ploughs of a suitable construction attached to the engine readily cast it to the right and left, and as there rarely falls a sufficient quantity in a single day to impede seriously the running of trains, there could be no great practical difficulty in keeping a line open for traffic if the railway track was placed on *an embankment* throughout its whole extent.

It is not possible, in a country like that between River du Loup and Truro, to find a line for a railway which would be free from cuttings; the surveys, indeed, indicate that some very heavy ones must be formed. It is, however, quite practicable with an increased outlay, to widen the cuttings and deepen the sides of them, so as to leave the rails elevated in the centre, in the manner shown in the accompanying sketch, and thus provide space sufficient within the slopes for the snow which the locomotives would throw off the rail track;—to form as it were a small embankment through the centre of each cutting. Thus by contriving to have the rails sufficiently elevated above the ground along each side, in cuttings as well as elsewhere, it is believed that it would be quite practicable to keep open the proposed railway in winter at a moderate cost.

By adopting a plan of construction such as suggested, and the drifts prevented in the manner already referred to, I can see no reason why trains should not be run between River du Loup and Halifax, with a higher degree of regularity than on the Grand Trunk Railway east of Montreal.

The sketch is intended to show a cutting with a rail track raised in the centre to afford an opportunity for throwing the snow easily into the space provided for it at each side.

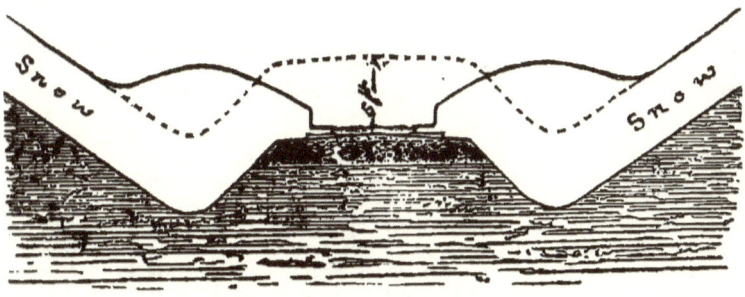

The snow is supposed to be five feet in perpendicular depth, the dotted line shows where its surface would be, supposing the Railway to have been closed all winter, and the full line shows where the snow would be deposited along the side, on being cast to the right and left from the rail track.

I see no other way of providing efficiently for the removal of the deep accumulation of snow which may be looked for in winter, particularly in the northern parts of the country, and therefore I consider it essential that a system of construction be adopted similar to that above described.

The increased width of cuttings required, will of course have the effect of swelling out the expenditure on the undertaking in the first instance; but this I consider unavoidable, as upon the means which may be furnished for facilitating the removal of snow, the regularity and consequent value of the Intercolonial Railway as a winter communication will mainly depend.

THE ESTIMATE OF PROBABLE COST.

In submitting estimates of the probable cost of the contemplated undertaking, it is necessary to allude briefly to the nature of the various services on which expenditure will be required. I shall therefore proceed to consider them in the order in which they properly come, viz:

1. *Engineering*, comprising all Exploratory, Preliminary and Locating Surveys. Designing, Inspecting and Superintending works;
2. *Right of Way and Fencing*;
3. *Clearing*;
4. *Permanent cottages for Workmen*;
5. *Telegraph*;
6. *Grading and Bridging*, comprising all the main works of construction in forming the Road-bed;
7. *Superstructure*, embracing Ballast, Ties, Rails and Rail-fastenings, for Main track and Sidings;
8. *Station Accommodation*, comprising all buildings and erections required for general traffic, for protection and repair of Rolling Stock, for wood and water services.
9. *Rolling Stock*;
10. *Contingencies*, including every possible expenditure directly connected with construction.

ENGINEERING.

The exploring, Surveying and Locating operations indispensable to the establishment of an undertaking such as that proposed, precede all other services and therefore the consideration of this branch of expenditure naturally comes first.

The surveys already made are not without their value, but a great deal has yet to be done before the location of any one line can be proceeded with. When it is considered that in a country so densely wooded as the one in question, where in much of it, a person under ordinary circumstances can scarcely see over fifty yards around him, in any direction except upwards, it will not be wondered at that the operation of determining in detail, the best position for a Line of Railway, is considered an exceedingly tedious and expensive matter.

In a level wooded country, or one with gently undulating slopes, it sometimes makes little difference in the cost of the work, or in the character of the gradients of a projected Railway, where the line is taken; and in such cases the first trial or random line through the woods, is not infrequently adopted for the Railway route with but slight modifications. In a country however whose features are characterized by great irregularities, and whose surface is covered with a dense vegetation, the information necessary to secure the best and least expensive location, can only be acquired by a series of laborious measurements.

A great deal of exploratory work will yet be necessary before the Intercolonial Railway can be proceeded with. It is in the highest degree important that the country should be thoroughly known and the best engineering route for the Railway fully and finally established before works of construction are commenced. It is always true economy to expend money on efficient surveys, and in this particular case vast sums may be wasted by an opposite course. The country is of such a character, more particularly in the Central and Northern districts, that almost any amount of money may be expended on a careless location; whilst sufficient time and attention bestowed on these preparatory services, would eventually prevent waste, disappointment and discredit. I consider it essential that ample provision be made in the estimate, for all the Exploratory and Surveying services referred to, as well as for the employment of an efficient professional staff in designing and superintending the proper execution of the miscellaneous works incidental to Railway construction.

RIGHT OF WAY AND FENCING.

The Province of Nova Scotia has in the construction of her Railways, instituted a system worthy of imitation, so far at least as the mode of providing the land on which to build them is concerned. Whilst the Railways are admitted to be for the general public good, it is justly assumed that the immediate locality through which they pass derives greater benefits from their construction than remote districts of the Province.

On the principle therefore that those who get the benefits should bear the burdens, the Legislature of Nova Scotia has enacted, that the several Counties intersected by the Railway, shall provide the "Right of Way" and bear the expense of separating it from the adjoining lands.

Of course the land is not taken from the owners without compensation, but the settlement of this question is left with the local authorities, and the amount of compensation, together with the cost of erecting fences, added thereto, is paid out of County funds and met by assessment in the usual way.*

This system is I believe readily acquiesced in by the people, those who do not happen to live in the counties through which the Railway passes, have no special "Right of Way" tax to pay; and those who have the tax to pay on account of their proximity to the line of Railway, consider themselves the most fortunate, as the trifling county charge is much more than counterbalanced by the great advantages secured.

In other respects the system adopted in Nova Scotia promises to result satisfactorily, the total expenditure on the Railway out of the Provincial Funds, will be reduced by the cost of Land Damages and Fencing; and the parties connected with its construction will not be required to resist exhorbitant claims too frequently made for alleged Land damages and which the local authorities can best adjust; and thus antagonism between the people of the Country and the Railway authorities will be avoided.

In the construction of the Intercolonial Railway there appears to be every reason why this system should be imitated, and I shall therefore in the estimate make no provision for the purchase of Right of Way, for Land Damages of any kind or for Fencing. Of course neither one nor the other will be required in those sections where the line may be built through unsettled Government lands. In cultivated districts only will the proposed arrangement be requisite and there it will have to be sanctioned by Legislative enactment.

CLEARING.

So soon as the preliminary and location surveys are completed, the clearing of the "right of way" may be proceeded with, on the line selected for the construction of the Railway.

The surveys will probably occupy the whole of the first year, but during this period it would be possible to complete the location of some sections earlier than others; in such sections the clearing might be proceeded with and this work may in part also be continued during the following winter, and thus allow the work of excavation to be commenced on the opening of spring.

The clearing ought to be made to a width of not less than three chains or about 200 feet for a threefold object: 1st. To remove all danger from trees falling across the rail-track; 2nd. To reduce the chances of injury to the track or to passing trains, by reason of

---

*The money payable for such lands and fencing shall form a county charge, but in the apportionment of the assessment the sessions shall have respect to the relative benefits derived from the railway by the several sections of the country, and shall apportion the assessment accordingly. *Chap.* 70, *Sec.* 24, *Revised Statutes of Nova Scotia*, 1864.

fires raging in the woods, a contingency not uncommon and frequently very troublesome in dry summers; 3rd. To allow space for the springing up of a second growth of spruce and other trees along each side of the Railway, which in a few years would attain a sufficient size to act as a natural and permanent snow-fence, should the adjoining lands become cleared of their timber.

BUILDINGS FOR WORKMEN.

On the completion of the Railway a large number of men will permanently be required upon its future maintenance. These men with their families will require a considerable number of cottage dwellings and tool houses. Such buildings should be regarded as necessary appendages to the Railway, and when so considered it would greatly facilitate the works of construction to have them erected in the first instance, of a permanent and suitable character; by permanent I do not mean expensive, comfortable log houses, warmly built, like the farm houses in Lower Canada and elsewhere, would serve every purpose.

These buildings should be provided along the line at about every five miles distance and at points convenient to good water. They ought to be proceeded with so soon as the exact position of the line is determined; they would during construction be serviceable as Engineers and Contractors offices and also as storehouses and dwellings. The outlay on them need not be great and I am satisfied it would be a profitable one.

A TELEGRAPH.

A Telegraph is now considered an indispensable adjunct to a Railway; it is essential to the proper and safe working of the line when completed, and therefore provision should be made in the estimate, for a fully appointed Telegraph, throughout the whole distance

Only those who have been engaged in Railway construction through districts remote from easy means of communication, will be able fully to appreciate the great advantages which would result from the possession of a line of telegraph, during the progress of works, through the roadless districts. A Telegraph, in all situations, is a convenience and a requisite of no little moment; but where ordinary means of communication do not exist, or exist only in the most primitive form, this modern and comparatively inexpensive means of conveying intelligence and directions would be doubly valuable. The importance of a Telegraph along the line of works during their progress would be so great that I am convinced its early erection would very favorably affect the expenditure on construction; and, as it must ultimately be provided, I would strongly recommend that it be furnished at the earliest practicable period, so soon, in fact, as it is possible to have the route cleared f its standing timber.

BRIDGING AND GRADING.

The various services above referred to, may be considered as preparatory operations to the commencement of the main works of construction. Surveying the country and laying out the line are of course indispensable preliminaries. The right of way must necessarily be secured. The clearing of the land must precede the erection of the Telegraph, and to some extent, also, the building of the cottages for workmen herein proposed, it would also open up a way for the taking in of men and supplies. Each service in its proper order would facilitate that which follows, and all that have been mentioned would either necessarily precede the works of excavation, grading and bridging, or render them less difficult of execution and consequently in proportion less expensive. .

Alll Bridges are intended to be built of durable materials and in the most substantial manner. Wherever it is practicable to cross a stream on an earthen embankment with an arch culvert for the water way, this system is preferred; but in cases where the height of the roadway above the stream is not sufficient for the introduction of arches, open beam culverts having walls of good masonry must be substituted.

All openings of more than twenty feet span, are intended to have wrought iron beams placed on substantial bridge masonry.

In establishing the Intercolonial Railway I think it would be mistaken and dangerous economy to introduce the construction of any bridge structures except those of a permanent and substantial character; and in determining the size of culverts and water courses, it will be important not only to make full provision for the passage of freshet water at the present day, but also to have in view an increased occasional discharge in the future, on account of the facilities for rapid drainage which the destruction of the forest and the cultivation of the land will afford.

With regard to the works of excavation and grading—for reasons already given, and mainly to facilitate the removal of snow from the track in winter, it is in contemplation to have the rails raised to a height, not usually adopted, above the adjoining surface of the ground. This will be more especially advisable throughout the northern portions of the country, where, in order to effect the object desired, it is proposed to avoid cuttings as much as possible; and when this cannot be accomplished, it is intended that the cuttings should be formed of sufficient width to afford space along each side of the track, for the snow to be cast by snow-ploughs.

Without some such provision as that above referred to, it is feared the cuttings would frequently be choked up with snow, during the winter season.

The quantities of excavation already submitted, have been computed on the assumption that the cuttings will be made to an avarage width of 30 feet at formation level, and with side slopes of one and a half horizontal to one perpendicular. It is, however, proposed to vary this width in actual construction, increasing it to 34 or even to 36 feet at points where on a better knowledge of the country and climate it is found the greatest amount of snow generally falls; at the same time making a corresponding decrease in the width, where the snow-fall is known to be on the average light.

Embankments are intended to be 18 feet in width at formation level with side slope of 1½ horizontal to 1 perpendicular; wherever embankments are exposed to the current of a stream, provision will be made for their protection by slope-walling.

In order to make the road-bed dry, firm and perfect, and also to reduce the difficulty and expense experienced in maintaining wet cuttings, it is proposed to adopt a system of thorough under-drainage, wherever the soil or sub-soil is at all wet.

THE SUPERSTRUCTURE.

Under this heading I shall embrace the Ballast, the Ties, the Rails, and the Rail-fastenings.

The Ballast is a most important element in the construction of a Railway and upon it greatly depends the durability of the Iron and the Rolling Stock. The best Railways, those which do the most business with the least outlay, are invariably found the best ballasted.

In many sections of the country between Truro and River du Loup, there are indications of abundance of material for Ballast, but as quality is more important than quantity, although a sufficiency of the latter is essential, care should be taken to have the very best, selected in the first instance, whatever it may cost. The estimate, which will shortly be submitted, provides for a quantity of 5,000 cubic yards per mile; this quantity if of good material, laid on a road-bed thoroughly drained, will, without doubt, make a good track, but less would scarcely be sufficient to accomplish the purposes of Ballast, in a satisfactory manner.

The cross-ties will be of the usual dimensions, made flat on two sides, six inches thick and nine feet long. The different kinds of timber available in various sections of the country for the making of Ties has already been referred to, the best which each locality can afford is intended to be employed.

With regard to the Rails and their fastenings, I would recommend a rather heavier pattern than has commonly been employed in this country, with the "fish" or some equally good splice joint.

In the estimate, I have allowed for the rail weighing with the joint fastenings 70 lbs. per lineal yard; on a Railway such as the one proposed, with heavy grades, and as a consequence, heavy Engines, I think this weight of rail, although costing more in the first place than a lighter one, will ultimately give greater satisfaction.

The joint fastenings are intended to be the most effective and reliable made, on account of the severity of the climate of this country.

The quality of the iron is of the utmost importance, and every care should be taken to secure the best manufactured. There is no economy in purchasing bad iron at a low price, as shipping, handling, transporting, laying and all other charges, are quite as much on inferior iron as on material of the best quality; besides which the durability of the one is so much greater than the other, that even if the best should cost considerably more originally, it will be found the cheapest in the end.

In the estimate an allowance of ten per cent. on the whole mileage of the Railway is made for sidings. It is believed that this proportion will be sufficient for operating the line until the traffic greatly increases.

STATION ACCOMMODATION.

With regard to the Station accommodation and general Depôt services, I would, in submitting an estimate of this kind, prefer defining the number of stations and character of buildings which in my opinion would be required. But as the route itself is quite an open question, it is impossible to judge what may be necessary, and therefore, I can only include in the estimate a uniform mileage charge for these services.

I may remark, however, that I consider an efficient water service with commodious wood-sheds, indispensable, and this should be the first thing looked to along the line.

With the exception of a few points where towns are touched and where proper accommodation must be provided, I can see no necessity for much expenditure on Station Buildings. Whilst I would strongly recommend that the Railway proper, and everything immediately apertaining thereto, such as Bridges, Culverts, Embankments, Ballast, Rails, &c., be made of the best materials and in the most substantial manner, so as to insure speed, safety and economy, in transit and maintenance; I think it would be unwise to expend money through the wilderness districts, in costly buildings, which for many years cannot be required.

If necessary let a fund be reserved for the purpose of being expended from time to time as required, and as traffic through the country gradually develops itself, but in the mean time, only a limited number of Station buildings, and these of the simplest character, need be erected.

Permanent establishments for the accommodation and repair of Rolling stock are indispensable; they will consist of engine stables, and workshops with machinery for repairs; they should be situated at such central and convenient points as may, on a full consideration of the advantages of each locality, be determined.

ROLLING STOCK.

It is difficult to form an estimate either of the kind or quantity of Rolling stock likely to be required, as so much depends on the character of the traffic, and this again is in a great measure governed by the route which may ultimately be selected.

I think that the best course is to provide a moderate quantity of Rolling Stock, comprising cars suitable for the different kinds of traffic; together with a reserve fund to be expended as the nature of the traffic develops itself and as increased equipment becomes necessary.

The Rolling Stock which I consider may with propriety be furnished in the first place, is in the following proportions:

15 Locomotives for every 100 miles of Railway.
 4 Sleeping Cars   "   "
 4 First Class Passenger Cars "   "
 8 Mail, Baggage 2nd Class Cars"
40 Box Freight Cars   "   "
80 Platform Cars   "   "
20 Hand Cars   "   "

These of the best description, together with a sufficient number of snow plows, either fitted to, or separate from the engine, can be furnished for $300,000, or at an average mileage cost of $3,000.

CONTINGENCIES.

In order to provide fully for every expenditure, it will be necessary to embrace in the estimate an allowance for contingencies, for miscellaneous expenses, and also a reserve fund for increasing the Rolling Stock as well as the Station accommodation.

There are various miscellaneous services which will be made a charge on the fund for contingencies, of which may be mentioned a telegraph, workmen's dwellings, road crossings in settlements, printing, advertising, &c. The estimate, would not be complete without embracing a fund for all these and other expenses incidental to construction. The allowance in the estimate, does not however provide for interest, discount, commission or other charges on capital.

THE ESTIMATES.

Having described in general terms the nature of the services for which expenditure of capital will be required, in the construction of the contemplated Intercolonial Railway, I shall now proceed to submit estimates of its probable cost. In doing so I may observe, that considering the character of the survey, no great pretensions to accuracy can reasonably be expected. At the same time I may add, that the knowledge I have acquired of the country by the recent examinations, induces me to believe that although the estimates are only approximations yet they need not under proper management be exceeded.

There are certain services which do not altogether depend on the measurements made on the lines of survey; on estimating the cost of these I deem it best to consider them uniform mileage charges. They are as follows:

UNIFORM MILEAGE CHARGES.

1. Clearing, Grubbing, Draining, &c................................$1,000 00
2. Superstructure, embracing Ballast, 5,000 cubic yards, Rails and joints, 70 lbs. per yard, Spike, Cross-ties, Tracklaying, and an allowance of 10 per cent additional for Sidings................10,500 00
3. Station accommodation... 1,000 00
4. Engineering.. 1,500 00
5. Rolling Stock... 3,000 00
6. Contingencies including miscellaneous services, and reserve fund for extra rolling-stock... 6,000 00

         Total...................... $23,000 00

Producing a total mileage charge of $23,000, which will be considered uniform throughout, and common to all lines.

In another place I have given the approximate quantities of excavation, masonry, iron, &c., required to complete the Grading and Bridging on various lines surveyed last summer.

I have computed these quantities at prices which I consider liberal and sufficient; the result is now embraced in the following Estimates:

1. TRURO TO MONCTON, NOVA SCOTIA DIVISION OF THE SURVEY.

Uniform Mileage charges above referred to, estimated 109
 miles at $23,000 per mile...................................... $2,507,000
Bridging and Grading, estimated from quantities deduced
 from exploratory survey.. $2,693,000
     Total estimate Truro and Moncton Division $5,200,000

2. RIVER DU LOUP TO APOHAQUI, NEW BRUNSWICK AND CANADA DIVISION OF THE SURVEY.

Uniform mileage charges 340 miles at $23,000 per mile $7,820,000
Bridging and Grading estimated from quantities deduced
 from exploratory survey....................................... $7,615,500
     Total estimate River du Loup to Apohaqui $15,435,500

        Grand Total...................... $20,635,500

This sum $20,635,500 is the estimate for the whole line by the route surveyed last summer, from River du Loup by way of the River Toledi, Green River and Gounamitz Valley, thence by Two Brooks, Wapskehegan, the upper waters of the Miramichi and Nashwaak, by the Keswick Valley and St. John River to opposite Fredericton, thence by the head of Grand Lake and Chowans Gulch to Apohaqui Station. It embraces also the section from the New Brunswick Railway to Truro in Nova Scotia.

This total sum divided by the length of line to be constructed, gives an average of very nearly $46,000 per mile.

I have already mentioned that the cuttings have been estimated to a uniform width of 30 feet at formation level, and explained that in actual construction it will be advisable to vary this width, in proportion to the average snow-fall at different points; towards the north the width should be increased while towards the south it may be decreased.

These contemplated changes although they need not affect the total cost of the whole line, will, of course, alter the proportion chargeable to each separate division, and thus the estimate for that part between Truro and Moncton, viz., $5,200,000 may hereafter be found in excess.

THE MATAPEDIA DIVISION.

An estimate may similarly be formed of that portion of the Bay Chaleur line, which was re-surveyed last summer, up the valley of the Matapedia, and in length 70 miles.

Uniform mileage charges as already estimated, 70 miles at $23,000 per mile..	$1,610,000
Bridging and Grading estimated from quantities ascertained from survey...	1,175,000
Total...........................	$2,785,000

The estimated cost of this 70 mile section is $2,785,000 including a mileage proportion of all the charges necessary to complete the line and put it in operation. The average cost per mile of this section is therefore $39,786, and as Major Robinson and Captain Henderson considered it the most formidable portion of the whole route, between Halifax and Quebec ; it would probably give a maximum and safe estimate of the cost of the route to which they refer, by applying this rate per mile to the distance yet to be constructed. Taking this course we have $19,853,214 as the total cost of the line between River du Loup, and Truro. Less than this total sum may suffice, but until the surveys are extended to all points where difficulties may probably exist. I do not think it would be at all safe to estimate the cost of the Bay Chaleur route (line No. 15) at a less sum than $20,000,000

With regard to the cost of the other lines mentioned in this Report, it is quite impossible for me without further surveys to judge, except by the simple rule of comparison. It has been shown that the average estimated cost per mile of the surveyed Central line, including all services and sufficient equipment, is very close on $16,000; and it has also been inferred, from a careful survey 70 miles in length, in the Matapedia District, that a line by the Bay Chaleur would cost $39,786, or in round numbers, $40,000 per mile. We can only assume, therefore, until better data is furnished that the other lines may cost an average rate per mile ranging from $40,000 to $16,000 ; it is even possible, judging from the knowledge I have acquired of the country, that some of the lines referred to, may cost a higher rate per mile than the latter sum.

In concluding this Report, I desire to express my obligations to those gentlemen whom I selected to assist me in carrying on the Surveys, but for the zeal and untiring energy which they at all times displayed, it would have been impossible for me to have completed so early and so easily the important service which the Government was pleased to place in my hands.

SANFORD FLEMING.
Civil Engineer.

APPENDIX A.

THE AGRICULTURAL CAPABILITIES OF NEW BRUNSWICK.

From a Report by Professor James F. W. Johnston, F.R.S., &c.

"Two very different impressions in regard to the Province of New Brunswick will be produced on the mind of the stranger, according as he contents himself with visiting the towns and inspecting the lands which lie along the seaboard, or ascends its rivers or penetrates by its numerous roads into the interior of its more central and northern Counties.

"In the former case, he will feel like the traveller who enters Sweden by the harbours of Stockholm and Gottenburg, or who sails among the rocks on the western coast of Norway. The naked cliffs or shelving shores of granite or other hardened rocks, and the unvarying pine forests, awaken in his mind ideas of hopeless desolatation, and poverty and barrenness appear necessarily to dwell within the iron-bound shores. I have myself a vivid recollection of the disheartening impression regarding the agricultural capabilities of Nova Scotia, which the first two days I spent in that Province, around the neighborhood of Halifax, conveyed to my mind. Had I returned to Europe without seeing other parts of that Province, I could have compared it only with the more unproductive and inhospitable portions of Scandinavia.

"A large proportion of the Europeans who visit New Brunswick see only the rocky regions which encircle the more frequented harbours of the Province. They must therefore carry away and convey to others, very unfavorable ideas, especially of its adaption to agricultural purposes.

"But on the other hand, if the stranger penetrate beyond the Atlantic shores of the Province, and travel through the interior, he will be struck by the number and beauty of its Rivers, by the fertility of its River Islands and Intervales, and by the great extent and excellent condition of its roads, and (upon the whole) of its numerous bridges. He will see boundless forests still unreclaimed, but will remark at the same time an amount of general progress and prosperous advancement, which, considering the recent settlement and small revenue of the Province, is really surprising. If he possess an agricultural eye, he may discover great defects in the practical husbandry of the Provincial farmer, while he remarks, at the same time, the healthy looks of their large families, and the apparently easy and independent condition in which they live."

The Agricultural capabilities of the Province as indicated by its Geological Structure.

"The Agricultural capabilities of a country depend essentially upon its Geological structure. That of adjoining countries also, especially of such as lie in certain known directions, may modify in a great degree the character of its soils. In reference to this vital interest of a State therefore, the possession of a good geological map is of much importance, not only as an aid in determining the cultural value of its own surface, of what it is capable, and how its capabilities are to be developed, but in throwing light also on the probable capabilities of adjoining districts. * * *

"An inspection of this map (No. 1,) shows that according to our present knowledge, the Province of New Brunswick consists mainly of five different classes of rocks, represented by as many different colours. The gray, which is by far the most extensive, represents the region of the coal measures, the crimson that of granite and mica slates, the brownish red that of the red sandstone, the pale blue that of the clay slates, the green that of the traps and porphyries, and the light purple that of the upper Silurian. The dark purple in the upper part of the map, represents the lower Silurian rocks, which occupy the northern region towards the shores of the St. Lawrence.

"I do not here enter into any details in regard to the order of superposition of these rocks, because that general order is fully detailed in books of Geology, because in this Province there are certain districts in which the local order of superposition is far from being determined, and because a knowledge of the order is by no means essential to a clear understanding of the relations of these rocks to the agricultural character of the soil which covers them.

"It is of more importance to understand—

"1. That rocks of all kinds are subject to be worn away, degraded, or made to crumble down, by various meteorological and mechanical agencies :

"2. That the fragments of the rocks when thus crumbled, form the sands, gravels and clays that usually cover the surface of a country, and upon which its soils are formed and rest; and

"3 That for the most part the materials of which the crumbled sands, gravels and soils consist, are derived from the rocks on which they rest, or from other rocks at no great distance. How they come to be derived occasionally from rocks at some distance, will be explained in the following chapter

"These facts show that a close relation most generally exists between the rocks of a country and the kind of soils which cover it. It is this relation which gives Geology its main interest and importance in relation to Agriculture.

"A. *The Coal Measures* which cover so large a breadth of New Brunswick, consist for the most part, of grey sand stones, sometimes dark and greenish, and sometimes of a pale yellow colour. The siliceous matter of which they consist, is cemented together or mixed with only a small proportion of clay, (decayed felspar principally,) so that when those rocks crumble, which they do readily, they form light soils, pale in colour, easily worked, little retentive of water, admitting of being easily ploughed in Spring and late in Autumn, but hungry, greedy of manure, liable to be burnt up in droughty summers, and less favorable for the production of successive crops of hay.

"Of course among the vast number of beds of varied thickness which come to the surface in different parts of this large area, there are many to which the above general description will not apply,—some which contain more clay and form stiffer soils—some which though green or gray internally, weather of a red colour, and form reddish soils, but lightness in texture and in colour forms the distinguishing characteristic of the soils of this formation. This single generalization, therefore, gives us already a clear idea of the prevailing physical characters of the soils over a large portion of the Province, and illustrates the nature of the broad views which makes the possession of Geological Maps so valuable to the student of general Agriculture.

"This coal measure district is further distinguished by the general flatness of its surface, undulating here and there indeed, and intersected by rivers, and occasional lakes, but consisting for the most part of table lands more or less elevated, over which forests, chiefly of soft wood, extend in every direction. These flat tracts are not unfrequently stony, covered with blocks of gray sandstone of various sizes, among which the trees grow luxuriantly, and from among which the settler may reap a first crop of corn, but which almost defy the labour of man to bring the land into a fit condition for the plough. Such land abounds, for example, behind Fredericton on the way to the Hanwell Settlement, and is scattered at intervals over the whole of this gray sandstone country.

"Another feature which results from this flatness is the occurrence of frequent bogs, swamps, carriboo plains and barrens. The waters which fall in rain, or accumulate from the melted snow, rest on the flat lands, fill the hollows, and, from want of an outlet, stagnate, and cause the growth of mosses and plants of various other kinds, to the growth of which such places are prop'tious. Thus bogs and barrens, more or less extensive, are produced. A comparison of the Geological Map (No. 1), with the Agricultural Map, No. 3, appended to this Report, will show that the greater number of the extensive barrens of this kind yet known in the Province, is situated upon this formation.

'The Miramichi, the St. John, the Richibucto, and numerous other rivers, run in part or in whole through this district. Along their banks a fringe of soil is often found better than the uplands present; and hence along the Rivers the first settlers found comparatively fertile tracts of country on which to fix their families and commence the earliest farming operations. The Intervales and Islands of the River St. John form some of the richest land in the Province; but this richness arises in a considerable degree from the circumstance that this River flows, in the upper part of its course, through geological formations of other kinds, and brings down from the rocks of which they consist, the finely-divided materials of which alluvial soils of the Counties of Sunbury and York for the most part consist.

"In other countries, as England and Scotland, the coal measures contain a greater variety of rocks than is found over the carboniferous area of New Brunswick. They are distinguished from the latter by frequent beds of dark-coloured shale of great thickness, which form cold, stiff, dark-coloured poor clay, hard to work, and until thoroughly drained, scarcely remunerating the farmer's labour. Numerous sandstones which occur among them produce poor, sandy and rocky soils, so that large portions of the Counties of Durham and Northumberland, in the north of England, long celebrated for their richness in coal, still remain among the least advanced, and least agriculturally productive of the less elevated parts of the Island.

"B. *The Upper Silurian Rocks*, coloured light purple, cover an extent of surface in New Brunswick only inferior to that formed by the coal measures. They form the northern portions of the Province, from the mouth of the Elmtree River on the East, and Jacksontown on the west, as far as the Canadian border. In other Counties these upper Silurian strata consist of various series of beds lying over each other, each of which gives rise to soils possessed of different agricultural values. This is particularly observable in the western part of the State of New York, where some of the richest soils are formed from, and rest upon, rocks of this formation. It is a matter of regret that in this Province the large extent of northern country, over which these rocks extend, has not been sufficiently explored to allow of such sub-divisions being traced and indicated on the Map. That they exist, I have seen reason to believe, in my tour through the country; but the time at our disposal did not allow Dr. Robb and myself to go out of our way to explore their character or limits.

"On this formation a large part of the richest upland soils of the Province are formed. The fertile, cultivated and equally promising wild lands of the Restigouche—and those on either side of the Upper Saint John, from Jacksontown to the grand Falls, rest upon, and are chiefly formed from the debris of these rocks, and were it not for the granite, trap, and red sandstone which intervene, similar good land would probably be found to stretch across and cover the whole northern part of the Province, from the Restigouche River to the region of the Tobique Lakes.

"From his published reports, Dr. Gesner had obviously collected much information regarding this region, which has hitherto been very difficult to explore; it would have cleared the way very much to an accurate estimate of its agricultural capabilities, had he been able by means of fossils or otherwise to establish the subdivisions among its several members which we believe to exist.

"The soils of this formation are for the most part of a heavier or stronger character than those of the coal formation. The rocks from which they are formed are generally slaty clays, more or less hard, but usually crumbling down into soils of considerable strength—as agriculturists express it—and sometimes of great tenacity. Among them also are beds of valuable limestone, more or less rich in characteristic fossils, and, so far as I am at present informed, chiefly from the reports of Dr. Gesner, the presence of lime in considerable quantity as an ingredient of the slaty rocks themselves—a chemical character of much importance—distinguishes the beds and soils of these upper Silurian rocks.

"A comparison of the Geological with the coloured Agricultural Map will shew that the pale red and blue colours which in the latter mark the position the first and second class upland soils, are spread over the same parts of the Province which in the former are coloured light purple—indicating the region of the Silurian deposits. Thus the geological indications and practical experience in these districts coincide. But the same comparison will shew that this concordance is by no means uniform, but that soils marked by the Nos. 3, 4, and even 5, occur upon parts of the country coloured upper Silurian in the Geological Map. This arises from one or other of several circumstances.

"1. From the defective state of our knowledge of the real geological structure of the interior part of the Province over which these rocks are supposed to extend. In the impassable state of the country there is a sufficient excuse for such knowledge being still incomplete. But the absence of such knowledge explains also why we cannot accurately describe and represent upon our Map the true relations of the Geology of large portions of this interior country to its practical agricultural value ; or

"2. To the fact that this formation, like that of the gray coal measures sandstone, has its level table lands on which water stagnates and produces extended barrens, and deep hollows in which swamps are formed, and burned lands, which the repeated passage of those devastating fires to which this Province has been occasionally subjected, has rendered apparently wortbless; or

" 3. To the proximity of trap and granite districts-(coloured green and carmine)-from which numerous blocks of stone and drifted gravel have been transported and spread over the Silurian surface so as to render the soils that rest upon it inferior in quality to what, according to the geological indications, they ought naturally to be.

" How much of the differences observable between the two Maps is due to each of these causes, can only be determined by future careful observation.

" C. *The Lower Silurian Rocks* occur abundantly in Canada East, forming the northern part of Gaspé and skirting the right shores of the St. Lawrence for a great distance. Like the upper Silurian strata they consist to a great extent of slaty rocks, more or less hard, and though not incapable of yielding rich soils, as is seen in the occasional productive valleys of Lower Canada, yet as they exist in New Brunswick they are covered for the most part with inferior soils.

The Agricultural capabilies of the Province, as indicated by practical Survey and examination of its Soils.

" D. *The Cambrian or Clay Slate Rocks*, coloured pale blue in the Geological Map, form two bands of which the limits are not well defined, running in a north-easterly direction across the middle of the Province, the more southerly of which bands doubles round the south-western extremity of the coal measures, or coal basin as it has been called, and forms part of Charlotte, St. John, and King's Counties. In nearly all countries these clay slate rocks are harder, less easily decomposed, and form more rocky and inhospitable regions than those of the Silurian formations generally. In this Province they do not change their general character, but they, nevertheless, as the Agricultural Map shews, are sometimes covered with soils of medium quality.

" The clay slates are for the most part formed like the Silurian strata, of beds of clay which have been gradually consolidated, but they are distinguished from the Silurian generally by two characters.

" First, by their greater hardness, which prevents their crumbling down and forming the close and often deep clay soils which the Silurian rocks occasionally yield. The clay slate soils, when freed from stones, are more of the character of what are called turnip and barley, than of wheat, oat and clover soils.

" Second, by their containing less lime than the Silurian rocks do. This is a character of great agricultural importance. In nearly every part of the world these Cambrian rocks are poor in lime. In climates suited to the production of peat they are also, from their impervious character, favorable to the formation of bogs. Hence in those parts of Europe where those slate rocks occupy areas of considerable breadth, draining and the use of lime are the first two measures of improvement by which the naturally unproductive agricultural qualities of these soils can be amended. The same means would probably prove profitable also on the clay slate soils of New Brunswick.

" E. *The Red Sandstones*. In Westmoreland, King's, Charlotte and Carleton Counties, a considerable breadth is coloured of reddish brown, designed to indicate the occurrence of these spots of red sandstone and red conglomerate more or less extensive. In regard to the exact position of these beds, whether they are all above or below the gray coal measures, or partly the one or partly the other, a question of great economical importance to this Province has been raised. As it chiefly refers however to the greater or less probability of obtaining coal, a point to which I shall refer particularly hereafter, and has comparatively little agricultural importance, I do not enter into the question here. A knowledge of the geographical position and extent of these beds is nevertheless of much importance, and it would be very desirable to have these both more exactly ascertained and more correctly delineated on the Map.

" The reason of this is, that the beds of which these red rocks consist, frequently crumble down into soils of great fertility. The richest lands and the best cultivated in Scotland rest upon such red rocks. It will be seen by a comparison of the Agricultural with the Geological Maps, that soils of first rate quality are known in this Province also, in Sussex Vale, in Sackville, on the Shepody River, and elsewhere, to occur in the neighbourhood of rocks of a similar character.

"The beds of these red sandstone formations consist—

"1st. Of red conglomerates which often crumble down into hungry gravels, producing good crops of oats and of grain when well treated, but having a disposition to "eat up all the dung, and drink up all the water."

"2nd. Of fine grained red sandstones, which crumble into red and sandy soils, light and easy to work, often fertile, and when well managed, capable of yielding good crops. They are such soils as the French inhabitants of this Province delight to possess, and of a large extent of such soils they are actual possessors.

"3rd. Of their beds of red clay, often called red marl, interstratified with beds of red sandstone, and crumbling down into soils which may vary from a fine red loam to a rich red clay. These are some of the most generally useful, and when thorough-drained, most valuable soils which occur among all our geological formations. In this Province these marls are usually associated with gypsum, as may be seen by the dots of brighter red which are here and there to be seen over the reddish brown portions of the Map. The soils may generally be calculated upon as likely to prove valuable for agricultural purposes wherever these beds of gypsum occur.

"Some of the sandstones of this formation, especially in the neighbourhood of beds of limestone, are themselves rich in lime. Thus a red sandstone collected in such a locality, three miles from Steves', in the direction of the Butternut Ridge, gave me upon analysis 17.31 per cent. of carbonate of lime, and 0.49 per cent. of gypsum. The crumbling of such rocks as this could hardly fail in aiding to fertilize the soil.

"The imperfect Geological Map of Dr. Gesner, which is lodged among the Records of the Land Office, and a more detailed copy of which is in the possession of the St. John Mechanics' Institute, represents the red rocks as much more extensive than they appear in the Map appended to this Report. One reason for this is, that he colours red the Parish of Bostford, and portions of the adjoining Parishes, where the red rocks do not appear, though the soils that cover the surface are red, and have evidently been derived from red rocks. This we observed in our recent tour through that country. On the Grand Lake also, Dr. Gesner colours red a considerable extent of country, upon which according to Dr. Robb, no true red rocks occur.

"Still these indications of Dr. Gesner, though not Geologically correct in a certain sense, are so in another sense, in which they are scarcely less useful to the agriculturalist. They indicate the general character of the loose materials that overlie the living rocks of the country and form its soils, and they tell more regarding those spots which is useful towards an estimate of its agricultural capabilities than a correct map of the rocks themselves would do. But the discordancies often observable between maps which exhibit only the characters of the rocks of a country, and those which exhibit its actual and experimental agricultural value, and the causes of such discordancies, will appear in the subsequent chapter.

"F. *The Granite, Gneiss, and Mica Slate*, coloured carmine, form a broad riband extending across the Province between the two bands of clay slate rocks. To the north of the slates also, and in the centre of the ungranted country, it forms a large patch of generally high land, the outlines and extent of which are by no means defined, and in the map are put down very much by guess.

"These regions are generally stony, often rocky and impossible to clear. When less stony, they sometimes give excellent soils after the less frequent rocky masses are removed, and in many places comparatively stoneless tracts of land occur on which clearances with less cost can readily be made.

"This description shews that the carmine regions are by no means agriculturally encouraging on the whole, judging by their geological character; but that they possess capabilities superior to those of the gray sandstone soils, is shewn by the experience of the farmers of these latter soils, that those fields generally turn out to be the best on which the granite boulders shew themselves most abundantly. The debris of the granite mixing with that of the sandstone rocks, improves its quality, gives it often more tenacity, and renders it more productive.

"The Agricultural Map will show that the soils along the carmine bands, and in the centre of the wild region between the St. John River and the Restigouche, though often very inferior, are not uniformly so. Were we better acquainted with the limits of the geological formations comprehended under this colour, we should be able, by means of them alone, both to form more accurate opinions in regard to the agricultural value of the several localities, and to represent them more correctly on geological maps, and to prescribe by mere inspection, the kind of ameliorations, mechanical or chemical, by which their natural qualities were likely to be improved.

"G. *The Trap-Rocks*, coloured green, which occur so abundantly among the southern clay slate and lower Silurian rocks, and in the wild country which forms the northern part of the Province, are the only remaining rocky masses which cover an extensive portion of the surface of New Brunswick. They form in this Province, a wild and generally a poor, rugged, rocky, inhospitable country. Lakes, swamps, and soft wood ridges, abound where they occur, and numerous blocks of stone try the patience and industry of the settler.

"Trap Rocks do not necessarily indicate the presence of unfertile soils. On the contrary, some of the most fertile spots in Scotland and England, are situate upon, and possess soils formed from these rocks. But such soils are formed only where the rocks are of a less hard and flinty nature, or at least are more subject to the degrading influence of atmospheric causes, and crumble to a soil readily. In such cases they generally form reddish soils of great richness, and when the soils are deep, it is found profitable to convey to some distance, and apply them as covering to less valuable fields.

"One cause of this fertility of trap soils is the large percentage of lime which these trap rocks frequently contain. This chemical character, for the most part, eminently distinguishes them from the granitic rocks, and indicates a very different mode of treatment for the sails formed from these two classes of rocks respectively.

"In New Brunswick, so far as my own observation goes, the trap rocks do not readily crumble, but remain hard and impenetrable by the weather, to a great extent. They do not usually, therefore, give rise to the rich soils which in many other places are formed from them. Hence St. John and Charlotte, partly owing to the less favorable clay slate and lower Silurian rocks which abound in them, partly to the obdurate trap, and partly to the numberless rocky masses which cover their surface, are justly considered among the least agriculturally promising Counties in the Province. I have witnessed, however, in both these Counties, that energy and determination can do much to overcome nature in New Brunswick, as well as in other parts of the world. Pleasing forms, and good crops, and comfortable circumstances, reward diligence and industry here in as wonderful a manner as in any other County in the Province.

"I do not dwell longer on this part of my subject. The general conclusions as to the agricultural capabilities of this Province which are to be drawn from the imperfect information as to its geological structure, which our Geological Map presents, are, on the whole, somewhat discouraging.

"The coal measures, the clay slates, the lower Silurian rocks, the granites, and the traps, are not, generally speaking, of a kind to give rise to soils of a fertile character, and these formations cover a large portion of the Province. The upper Silurian and red sandstone formations, on the other hand, promise much agricultural capability, and soils prolific in corn; and they also extend over a very considerable area. Were the geological exploration more complete, our deductions from this source of information would be more precise, more to be depended on, and possibly also more favorable, for reasons which will in some measure appear from what has been already stated. It is to be hoped that Your Excellency, and the Houses of the Legislature, will see the propriety, at an early period, of resuming this important exploration.

"More detailed and positive conclusions as to the absolute and comparative values of the soils in the different parts of the Province, on the different geological formations, and on the different parts of the same formation, the sub-divisions of which, as I have said, have not been made out, will be arrived at by means of the practical survey which forms the subject of the next Chapter.

"Although the geological structure of a country throws much general light on the geographical position, on the physical and chemical characters, and on the agricultural capabilities of the soil of a country, it does not indicate—

"1st. The absolute worth or productiveness of the soils in terms of any given crop— as that the red sandstone soil would produce so many bushels of wheat, or the clay slate soil so many of oats; &c.—

"2d. Their relative productive powers when compared with each other—as that if the coal measure soils produce twenty bushels of any grain, the upper Silurian would produce thirty bushels.

"Such absolute and relative values can only be ascertained by an actual trial and experience of absolute fertility of the soils in some spots at least, and by the personal inspection and comparison of the apparent qualities, with what is known of the origin, the composition, and the absolute productiveness of each.

"Again, the geographical *limits* of the several formations, as represented in the Geological Map, do not precisely indicate the limits of the several qualities of the soils which are naturally produced from them. The *débris* of one class of rocks frequently overlap the edges, and sometimes cover a considerable portion of the surface of another class of rocks adjoining them, in a particular direction, and thus cause the soils which rest upon the latter to be very different from what the colours of the Geological map would lead us to expect.

"In this country it is observed that the fragments of the different formations have very generally been drifted from the north or north-east to the south or south-west, probably by some ancient current similar to that which now brings icebergs from the polar regions, and which took its direction across this part of North America when it was still beneath the level of the sea. Hence the surface of one rock, or the debris derived from it, is very apt to be covered by a layer of a different kind, derived from rocks which lay at a greater or less distance towards the north or north east.

"This is most easily seen in the case of the red sandstone rocks, the debris of which, when drifted over the adjoining formations, impart a different colour to the soils which rest upon them. Thus on ascending the Tobique two or three miles above the Narrows, on the right bank of the River, a layer of red drift, a few feet in thickness, derived most probably from the red rocks above the rapids, is seen to rest on a thick bed of slate drift, and to form the available surface. Similar red drift extends itself in a similar direction from the red rocks of Sussex Vale; and Dr. Gesner, in his interesting reports, describes similar drift as visible along the shores of Grand Lake, and in many other localities.

"Sometimes, also, the upper rocks, which formerly overspread the surface of a country, have been worn down, washed away, and entirely drifted off, leaving us only the power of inferring that they once existed by the layers of fine mud, sand or gravel derived from them, which we observed upon the lower rocks which still remain.

"This is seen in New Bandon Parish, where the red soils appear to be chiefly derived from red rocks, which formerly existed in the direction of the Bay of Chaleurs; and in the Parish of Botsford, in Westmoreland County, the fine red soils of which have been drifted from Prince Edward Island, or from rocks in that direction, which have now disappeared.

"Further, it not unfrequently happens that the drifted materials which cover the surface of a country, and which form its soils, consist of the debris of two or more entirely different kinds of rock mixed together, as we readily understand that such different materials might be mixed together, if the same current were to pass, as the River St. John does, in succession over a series of different geological formations, and to mingle together in the same sea bottom, and in different proportions, the fragments of all. The nature of the soil thus formed would not be indicated either by that of the rock on which it rests, or by that of any one of the ten or more rocks from which it had been partially derived. Thus while an intimate relation undoubtedly does exist between the soils and rocks of a country in general, and a very special relation between any given soil and the rock from which it has been derived, so that the inspection of a Geological map will convey to the instructed eye a true general notion of the agricultural character and capabilities of the country it represents, still it does not exhibit to the eye, as I have said, the absolute and comparative fertility of its different soils in terms of any given crop, nor can it, in a country like this, precisely define the limits which separate soils of one quality from those of another.

"These points are only to be ascertained by special inquiry, and by a special survey and personal inspection. To make such inquiries and such a personal inspection, was among the main objects of my tour through the Province. The results of what I saw and learned myself, together with much other information obtained from the documents contained in the Land Office, from Dr. Gesner's Reports, and from other sources, I have been able, chiefly through the indefatigable and most willing assistance lent to me by Mr. Brown, to embody in the maps No. II. and No. III., attached to the present Report.

"In these maps I have represented by different colours and figures, the different qualities of soil in the Province, and the geographical position and approximate extent of each quality. For this purpose I have divided the soils into five different qualities, represented by a series of numbers, of which No. 1 indicates the best and No. 5 the worst quality.

"The special varieties of soil denoted by the figures and numbers, are as follows:—

"No. I. on the uncoloured, and the bright red on the coloured map, denote the soil of the best quality in the Province. This consists chiefly of river intervales, islands, and marsh lands. It is only of limited extent, and is confined, for the most part, to the course of the River St. John, that of the Petitcodiac, and to the neighborhood of Sackville.

"No. II. and the pale red colour, denote the best quality of upland, and such portions of good intervale and marsh land as are not included under No. I. It is to be understood, however, that there is much marsh land, both dyked and undyked, which does not deserve a place even under this second head. This first class upland exists chiefly in the Counties of Carleton and Restigouche.

"No. III. coloured blue, is the second rate upland, inferior to No. II., but still very good in quality. It represents the medium soils of the Province, and stretches over a much larger surface than any of the other colours.

"No. IV., coloured bright yellow, is inferior in quality to any of the others. It is decidedly inferior or poor land, resembling the least productive of that which is now under cultivation. It consists for the most part of light sandy or gravelly soils, hungry, but easily worked, or of stony and rocky ground, which is difficult and expensive to clear, but in some parts of Charlotte County, productive when cleared.

"This class also includes lands covered with heavy hemlock, and other soft wood, which, though hard to clear, and unfavorable for first crops, may hereafter prove productive when it has been submitted fairly to the plough. It will be seen that a great extent of this bright yellow land exists in the northern half of the Province.

"No. V. coloured pale yellow, includes all which in its present condition appears incapable of cultivation.

"The naked flats, distinguished as bogs, heaths, barrens, carriboo plains, &c., are all comprehended under this colour, and tracks of swampy country, which at present are not only useless in themselves, but a source of injury to the adjoining districts. All this pale yellow is not to be considered absolutely irreclaimable, but to be unfit for present culture or for settlement, till much larger progress has been made in the general improvement of the Province. The dark spots, coloured with Indian ink, represent the localities of some of the naked and barren plains which are included under this No V.

"It is not to be supposed that I or my travelling companions have been able to inspect, even cursorily, the whole of the country we have thus ventured to colour, and to distinguish by numbers. The country we have actually seen and explored during our late tour may be judged of from the green lines traced on both maps, which represent the route we took, and the country we actually went over. Our knowledge of the rest has been gathered from numerous persons whom we met with in different parts of the Province, from the reports and surveys deposited in the Land Office, and from observations of Dr. Gesner. Though far from being correct, these maps are valuable, both as an approximation to the truth, and as embodying nearly all that is at present known as to the soils of the Province. Your Excellency will, I am sure, both be inclined to value them more, and to make larger allowances for their want of correctness, when I mention they are the only maps of the kind of any country which, so far as I know, have yet been attempted, and that they have been of necessity executed in a very short period of time for so extensive a work.

"The relative areas, or extent of surface covered by these several soils, as they are represented in the coloured map, are very nearly as follows :

    No. I. coloured bright red,.. 50,000 acres.
    No. II. coloured light red,... 1,000,000 "
    No. III. coloured blue,................ .... 6,950,000 "
    No. IV. coloured bright yellow,................................. 5,000,000 "
    No. V. coloured pale yellow,...................................... 5,000,000 "

       Total area of the Province,........................ 18,000,000 acres.

" The area of the Province has been calculated so as to include the territory within the boundary, as it may possibly be determined, between New Brunswick and Canada.

" Such are the relative geographical limits of the soils of different qualities in the Province, and the areas covered by each respectively, according to the best information I have been able to collect.

" The absolute values of each variety of soils in terms of the staple crops of the Province, I have estimated as follows :—

" It is usual to talk and judge of the absolute or comparative value of land in New Brunswick by the quantity of hay it is capable of producing. I have taken this crop therefore as one standard by which to fix the absolute and relative value of the different qualities of the soil in the Province. Then of the grain crops—oats, taking the whole Province together, is the most certain, and probably the best in quality. The culture of the oat is extending also, and the consumption of oatmeal as a common food of the people, is greatly on the increase. I take this crop therefore as a second standard. I assume also, but this is an arbitrary assumption, that as an index of the value of land at this time in this Province, with its present modes of culture, 20 bushels of oats are equal to a ton of hay. In other words, I assume that where a ton of hay can be produced, twenty bushels of oats may be produced, or *its equivalent of some other variety of human food*.

" Thus I have the means of giving a value to the different varieties of soil, in terms either of food for stock or food for man.

" I have classified the soils of the Province, therefore, in terms of these crops at the following absolute and relative value per imperial acre.

No. I.   will produce 2½ tons of hay, or 50 bushels oats per acre.
No. II.   "   2 tons   "   40 bushels   "
No. III.   "   1½ tons   "   30 bushels   "
No. IV.   "   1 ton   "   20 bushels   "

" The only reasonable objection which so far as I know can be made against this estimate is, to the value in oats assigned to the quality of the soils called No. 1.

* " It may be correct to object that this first class soil does not in practice produce 50 bushels of oats, but the real effect of this objection is very small : First, because nearly all this land is yearly cut for hay : Second, because grain crops (except in Sunbury, the Indian Corn,) do not succeed upon it in consequence of their rankness, which makes them lodge and refuse to ripen : and, Thirdly, because under proper culture in this climate, land that produces 2½ to 4 tons of hay, as the first class intervale and dyked marsh does, *ought* also to bear easily and to ripen upwards of 50 or 60 bushels of oats.

" The whole production of food for man or beast which the Province would yield, supposing all the available land to be cultivated according to the present methods, and that hay and oats bear to each other the relation of one ton to twenty bushels, would therefore be—

	Tons of Hay.	Bushels of Oats.
1st Class,..	125,000 or	2,500,000
2nd Class,.......................................	2,000,000 or	40,000,000
3rd Class,..	10,425,000 or	208,500,000
4th Class,..	500,000 or	10,000,000
Total produce,.........................	13,050,000	261,000,000

Being an average produce per acre over the thirteen millions of acres of available land, of 1½ tons of hay or 27 bushels of oats.

" What amount of population will this quantity of food sustain ?

" There are various ways by which we may arrive at an approximation to the number of people which a country will comfortably maintain upon its own agricultural resources. The simplest and the most commonly adopted in regard to a new country like this, is to say, if so many acres now in cultivation support the present population, then, as many times as this number of acres is contained in the whole available area of the country, so many times may the population be increased without exceeding the ability of the country to sustain it.

" Thus in New Brunswick, there are said to be at present about 600,000 acres under culture, and the produce of these acres sustains, of—

Men, women and children	210,000
Horses and cattle	150,000
Sheep and pigs	250,000

"But 600,000 are contained in 13,000,000, the number of available acres in the Province, nearly 22 times, so that supposing every 600,000 acres to support an equal population, the Province ought to be capable of feeding about:—

Men, women and children	4,620,000
Horses and cattle	3,300,000
Sheep and pigs	5,500,000

The human population and the stock maintaining the same relative proportions as they do at present.

"But this estimate is obviously only a mere guess, and by accident only can be near the truth, because supposing the quantity of land actually in culture to be correctly stated, (which cannot with any degree of confidence be affirmed,) the important consideration is entirely neglected, that the land now in cultivation may be much superior in quality to those which are in a wilderness state. This indeed is very likely to be the case, as the history of agriculture shows that the least productive lands by nature, unless they are much more easy to work, are always the last to be brought into cultivation. It leaves out of view also the question of fuel, which we shall by and by see has a most important relation to the agricultural capabilities of a country and its power of supporting a given amount of population.

"But from the date above given we can approximate to the truth in another way, answering directly the question, what amount of population will the produce we suppose the Province able to yield, maintain?

"If we suppose a full grown man to live entirely upon oats without other food, he will require to support him for twelve months, about 1000lb. of oatmeal, equal to about 2000lb. of oats, which at the low average of 35lb. per bushel, amounts to 57 bushels. If we allow that each of the population, big and little, consumes 40 bushels, an apparently high average, then the consumption of each individual, according to our estimate of the comparative productive powers of the land, in regard to hay and oats, would be equivalent to two tons of hay, in other words, the breadth of land which would grow two tons of hay would on an average support one individual if fed upon oatmeal.

"The usual allowance for the winter feed of a horse in this Province is four tons of hay, and for a cow two tons, sheep and pigs may be estimated at a quarter of a ton each.

"The cattle and horses together are estimated at 150,000. If the relative proportions of the two kinds of stock be as in Canada West, about four to one, then the entire population and live stock, (poultry, dogs, &c., &c., excluded,) would require for their support the following amount of produce, calculated in tons of hay:

210,000 at 2 tons each	420,000	tons.
30,000 horses, 4 tons each	120,000	"
120,000 cattle, 2 tons	240,000	"
250,000 sheep and pigs, ¼ ton	62,500	"
	842,500	

"But we have seen that the average produce in hay of the whole 13,000,000 acres of available land may be estimated at one and a third tons per acre,—the above 842,500 tons of hay therefore represent 631,875 acres of land of average quality.

"It will be observed that this sum comes very near the extent of land supposed to be at present actually cultivated in the Province. It is also about one-twentieth part of the whole available area (13,000,000 acres) in hay; so that the Province, according to this mode of calculation, be supposed capable of supporting twenty times its present number of inhabitants and of live stock, that is—

Men, women and children	4,200,000
Horses	600,000
Cattle	2,400,000
Sheep and pigs	5,000,000

" If the proportion of animals materially diminish, of course the number of human beings which the country is able to support, would proportionably increase.

" Those who are familiar with the feeding of stock will have observed that in the preceding calculation I have allowed for the support of the live stock only during the seven months of winter, and that no land has been assigned for pasture during the remainder of the year while the hay is growing.

" It will be also observed, however, that I have supposed all the stock to be full grown, and have assigned a full allowance of hay to every animal, whatever its age. A considerable surplus, therefore, will remain unconsumed when the winter ends, which will go some length in feeding the stock in summer, or, which would be preferred, in allowing land to be set aside for pasture or for soiling the animals with green food in the stables.

" Again, by referring to the relative proportions of land employed in raising food for the human and the animal population, in the relative numbers in which they exist in New Brunswick, as they are given in a preceding page, it will be seen that about equal quantities are devoted to each. That is to say, that nearly half of the land will always be under a grain culture, and will consequently be producing a large quantity of straw of various kinds, upon which all the stock will be more or less fed.

" I do not stay here to remark on the unthrift which I in many parts of the Province observed, in the use of straw from different grains, nor upon the greater good which might be derived from this part of the crops, under a more skilful mode of feeding. I only observe that the two indefinite allowances above made will, in my opinion, amply make up in the whole for the additional quantity of food necessary to maintain the stock during the summer months over and above the quantity of hay adopted in my calculation

" Before quitting the general question as to the food which the land will raise, and the population it will support, there are two additional observations which it is necessary to introduce.

" First.—That I have made no allowance for the human food produced in the form of beef, mutton, pork, milk, cheese and butter. The hay grown on the one half of the surface of the country is, for the most part, consumed in the manufacture of these articles. When a calculation is made of the quantity of human food raised in this way, the numerical rate of the sheep and pigs to the human population being taken as it is in this Province at present, and the dead weight of the stock at the average which the common breeds usually attain by the present system of feeding, it appears that the beef, mutton, pork, and milk, ought alone to support a population, equal to about one-third of that which the corn land sustains.

" Thus the whole capabilities of the soil in respect to the support of population, may be represented by—

Men, women and children...5,600,000
Horses.. 600,000
Cattle...2,400,000
Sheep and pigs.. 500,000

" Second.—That I have made no reference to the Fisheries which are already so large a source of wealth to the Province, and of food to the people. The value of this supply of food may be allowed to stand against and pay for the West India produce, and other necessaries of life which they cannot raise themselves, but which in addition to their beef, milk and meal, the inhabitants will require.

" That we appear to fix at upwards of five and a half millions the amount of population which New Brunswick, according to the data we have before us, would in ordinary seasons easily sustain. But here the question of fuel comes in to modify in a more or less remarkable manner our calculations and opinions upon this important subject. This question is deserving of a separate consideration.

Actual and comparative productiveness of the Province, as shown by the average quantities of Wheat and other Crops now raised from an Imperial acre of Land, in the different Counties.

" In the preceding I have given a sketch of the general agricultural capabilities of New Brunswick, as they may be inferred from its geological structure, and of the absolute and comparative productive qualities of its soils, as deducted from practical observation and inquiry. But the natural qualities of the soil may be neglected, overlooked, or abused. The actual yield of the land may be very disproportionate to its possible yield. The crops may be less than they ought to be, for one or other of many reasons, to which I shall advert in the subsequent part of this Report.

"It is in fact the actual condition of the practical agriculture in the Province which will determine the actual productiveness of its soils ; while on the other hand, the possible productiveness of its so being known, the amount of produce actually raised will serve as an index or measure of the actual condition of the agricultural practice.

"Looking at the matter in this point of view, it appeared to me of much consequence to collect as widely as could be done with the time and means at my disposal, numerical statements as to the actual number of bushels of the different kinds of grain and root crops usually cultivated within the Province, which were now raised from an imperial acre of land in its several Counties. Finding it impossible to collect all these data myself, I addressed a Circular to the farming proprietors and Agricultural Societies in the several parts of the Province, and from the answers I have received, the Tables (Nos. IV. and V.) have been compiled. They are not to be considered as rigorously accurate ; they are liable to certain suspicions to which I shall presently advert; but they are the first of the kind that have ever been compiled in reference to this Province ; the numbers they contain have been given, I believe, according to the most careful judgment of the persons by whose names they are guaranteed, and in the absence of better information, they are deserving of a considerable amount of credit.

"These Tables exhibit several facts of an interesting and some of a very striking kind ; thus—

"1. *The produce actually raised differs much in different parts of the same County.* Thus, in Westmoreland, one person returns 15 and another 20 bushels as the average produce of wheat; in King's, one gives 15, another 25 ; in Sunbury, one gives 12½ and another 20 ; in York one gives 15 and another 32, and so on. Similar differences exist in regard to other kinds of grain.

"Such differences are natural enough, and do not necessarily imply any incorrectness in the several returns. They may arise from natural and original differences in the nature of the soil; from its being more or less exhausted by previous treatment; or from the actual farming being in one case better than in another.

"2. *In regard to Wheat,* the lowest minimum is in Queen's, where 8 bushels are given as sometimes reaped. In St. John, Charlotte and King's, the minimum is 10 bushels; from Carleton no return is given, and altogether the answers from that County are few and therefore defective. The largest maxima are from Kent, Charlotte and York, where 40 36 and 32 bushels respectively are sometimes reaped.

"3. *In regard to Oats,* only one County (Queen's) ever reaps less than 25 bushels an acre, according to these returns. In that County, as little as 13 bushels is occasionally reaped.

"In four Counties the crop sometimes reaches 60 bushels ; in two others, 50 ; in one, 45 ; and in four, to 40 bushels an acre. These numbers indicate what is indeed confirmed by numerous other circumstances, that not only do oats succeed admirably, but that they are well adapted to, and are one of the surest or least uncertain crops now grown in the Province.

"4. *As to Maize or Indian Corn,* it will be seen that only in two Counties, (King's and Queen's,) is the minimum stated at less than 35 bushels an acre, while in four counties, the smallest yield of this crop is represented at 40 and 45 bushels. In Sunbury, the large return of 80 bushels an acre is sometimes obtained, and in Charlotte and Northumberland, as much as 60 bushels.

"This crop is liable to injury from early frosts, and is therefore somewhat uncertain in this climate, which by the great heat of its summers is otherwise well adapted to its growth. The four Counties of Sunbury, Queen's, Charlotte and Northumberland, would seem by the returns to be specially favorable to this crop. If so, its larger cultivation should be encouraged.

"5. *As to Buckwheat,* 15 bushels an acre are the smallest return, while crops of 70 bushels are sometimes reaped. The experience of the last two years has shown not only that this crop in one or other of its varieties is tolerably certain, but that it is well adapted to the exhausted condition of many of the soils, and affords also a very palatable food.

"6. *Of Potatoes*, the smallest return is 100 bushels, or about three tons per acre; but in Queen's County, a thousand bushels, about fourteen tons, are sometimes obtained. This latter amount is rarely surpassed, even in the west of Scotland, the north-western parts of England, and in Ireland, where the soil and climate are most propitious to this root.

"7. But the most striking fact brought out by these Tables is the comparative high number by which the average produce of each crop in the entire Province is represented. These averages appear in the last line of the second table, and are as follows:—

VI. Wheat......................19 11-12, say 20 bushels.
    Barley......................29 bushels.
    Oats........................34 do
    Buckwheat...................33¾ do
    Rye.........................20¼ do
    Indian Corn.................41¾ do
    Potatoes....................226½ do or 6½ tons.
    Turnips.....................456 do or 13½ tons.

" No very correct or trustworthy averages of the produce of the different crops in England, Scotland, or Great Britain, generally, have yet been compiled. It is believed, however, that 25 bushels of wheat per imperial acre, is a full average yield of all the land in Great Britain on which this crop is grown: some places, it is true, yield from 40 to 50, but others yield only 10 or 12 bushels per acre.

" It is of less importance, however, to compare the above averages with any similar averages from Europe. It will be more interesting to Your Excellency and the Legislature, to compare them with similar averages collected in other parts of the Continent of America.

" In the yearly volume of the transactions of the New York State Agricultural Society, for 1845, an estimate is given of the produce per imperial acre of each kind of crop in the several Counties, and a series of general averages for the whole State. The State averages, compared with those for New Brunswick above given, are as follow:—

VII.   *Average produce per Imperial Acre.*
                    State of New York.   New Brunswick.
Wheat............... 14 bushels   20 bushels.
Barley.............. 16 "   29 "
Oats................ 26 "   34 "
Rye................. 9½ "   20½ "
Buckwheat........... 14 "   33¾ "
Indian Corn......... 25 "   41¾ "
Potatoes............ 90 "   226 "
Turnips............. 88 "   460 "
Hay.................   "   1¾ tons.

" The superior productiveness of the soils of New Brunswick, as it is represented in the second of the above columns, is very striking. The irresistible conclusion to be drawn from it appears to be, that looking only to what the soils under existing circumstances and methods of culture are said to produce, the Province of New Brunswick is greatly superior as a farming country to the State of New York.

APPENDIX B.

AGRICULTURAL CAPABILITIES OF THE MATAPEDIA DISTRICT.*

" The Township of Restigouche is situated at the head of the tideway on the Restigouche, which forms its southern boundary ; it is divided from the township of Matapedia by the river of that name, up which they extend ; its general character is an elevated table land, from two to eight hundred feet above the sea; the surface is much broken with ravines and narrow valleys, the sides of which often form angles with the horizon of from twenty to forty degrees ; the summits of the hills are of considerable extent, presenting in some cases an even surface for several miles in length, by upwards of half a mile in width. The ground is a brownish or yellow loam, of a good quality, free from stones, the substrata being generally trap rock, which when decomposed forms an extremely fertile soil. It is well timbered with yellow and brown birch, maple, white birch, balsam, fir, spruce, beech and rowan tree or mountain ash ; the latter named woods, intermixed with white pine and cedar, also prevail on the sides of the hills, which, from their excessive steepness, do not occupy as much room as might be expected from the broken appearance of the ground ; the extent of the flats in the ravines and valleys is limited ; the timber on these places is chiefly soft wood, with some ash and elm.

" The description above will apply to the Township of Matapedia, which is also bounded on the south by the Restigouche. Limestone exists in both these Townships, sufficient for building purposes and manure whenever it may be required ; the ground is well supplied with springs and small brooks, the water of which is of a good quality.

It might be supposed, that from its elevation, the tract of country just described, would, in a great measure, be unfit for cultivation ; the crops raised, however, in this district, at the height of a thousand feet above the sea, ripen as early, return as much, and are of as good quality as those grown in the valleys.

" A few years ago the country around the Bay of Chaleurs was considered unfit for raising wheat ; experience has proved this unfounded, and it now produces all the kinds of grain raised in Eastern Canada. The climate does not appear colder than in the district of Quebec. Fogs are little known. Showers of snow fall about the end of October ; winter generally sets in, in the middle of November, but fine weather often continues to the end of the month ; the average height of the snow is four to five feet when deepest ; it disappears about the beginning of May, and the ground is fit for sowing a few days afterwards.

"Owing to the direction of the Baie des Chaleurs and River Restigouche, the winds are either westerly or from the east ; strong gales are of rare occurrence.

" The well cultivated grounds in the neighbourhood of Dalhousie, yield, of wheat, thirty to thirty-two bushels per acre; peas, about the same ; oats, forty to forty-eight ; barley, forty-five to sixty ; potatoes, three to four hundred ; carrots, two hundred and seventy to three hundred bushels per acre ; hay, two to four tons per acre. The weight of grain exhibited at the Agricultural Shows in the district, has been as follows : spring wheat per Winchester bushel, sixty-four to sixty-seven pounds ; fall ditto, sixty-six ; Siberian wheat, sixty-four to sixty-five ; oats, forty-two to forty-eight and a half ; barley, fifty-four to fifty-six ; field peas, sixty-six to sixty-seven pounds.

" On new land, not cleared of stumps, the yield of wheat has been thirty to one ; fifteen to twenty to one is not unusual. * * * *

" Two thirds of the surface of these townships, (Restigouche and Matapedia,) is of the quality already described, and comprise an area of nearly one hundred thousand acres of excellent land, that is from the Restigouche to Clark's Brook on the east side, and Mill Stream on the west side of the Matapedia.

" On the east side of the Matapedia from Clark's Brook the appearance of the country is extremely unfavorable ; steep hills rising from the river's edge, in many places denuded

---

*Report to the Honorable the Commissioner of Crown Lands, by A. W. Sims, November, 1848.

of wood by fire, and in others covered with a close growth of soft wood; the soil in general shallow and full of small stones. Of this section eleven miles in length by five broad, not more than an eleventh or five thousand acres is fit for cultivation."

"The aspect on the west from the river is not much different from that of the other side; the ground, however, though much broken by ravines is of a better description, the fires have done less damage to the timber which is a mixture of hard and soft wood. About half of the ground between Mill Stream and McKennon's Brook, embracing an extent of twenty-eight square miles, may be considered capable of advantageous cultivation; this would give nine thousand acres; it is well watered by the brook just mentioned and by that known as Connor's Gulch. Continuing on the west side of the river above McKennon's Brook, the surface in general is of less elevation than in the country already described; moist ground is more frequent, the timber consists of balsam fir, spruce, yellow, white and black birch, maple, cedar and white pine; in swampy places cedar and black and grey spruce predominate. The soil though much inferior to that at the mouth of the Matapedia, may be considered as of a fair quality; this will apply generally to the foot of the lesser Lake Matapedia, embracing an extent of eighty miles. About two-fifths or twenty thousand acres may be considered good."

"On the east side from Pitt's Brook, and across the Casapscul to near Fraser's Brook, the soil and timber is of the same description as on the other side, the ground is drier, and but few maple trees are found, fires have destroyed a great portion of the wood near the Matapedia, raspberry and other bushes, small white birch and poplar are now found in these places."

"Twenty thousand acres or about half of this section may be considered good land."

"Between Fraser's Brook and Fifty-six mile Brook near the southern boundary of the Seigniory of Matapedia, the soil, timber and character of the soil is diversified; from Fraser's Brook to the head of Little Lake the ground is in general very strong, rough and broken; a portion, however, is fit for cultivation near the shore, and after reaching the summit of the ridge which does not extend more than from three-quarters to a mile back, the soil improves and is covered with a good growth of fir, white, yellow and black birch, maple, cedar and white pine, and the general elevation of the ground is not much over two hundred feet, excepting one or two hills. From Little Lake to Fifty-six mile Brook there are flats bordering on the river, well timbered and sometimes of considerable extent."

"The available ground on this section which exceeds forty-five square miles, will amount to about half of its extent, fifteen thousand acres."

"On the west side of Little Lake and to the Seigniory of Matapedia, the general character of the soil and timber does not differ essentially from that of the section just described. At the base and partly up the sides of a hill near the foot of the Lake, (rising six or seven hundred above it) the timber is chiefly maple and other hard woods, the flat bordering the river is wider than in other places, the interval formed by alluvial deposits also extends up the Umqui, the mouth of which is near the Seigniorial line; ash, elm and the timber already mentioned as predominating in this district cover these places."

"The ground fit for cultivation in this section, forty-eight square miles in extent, is about seventeen thousand acres."

"The Seigniory Matapedia extends a league round the lake, and contains about ninety thousand acres in superficies; near the southern end of the lake there is a chain of hills bearing south ten degrees west nearly a thousand feet high, with a base from three to four miles broad; around the foot, and for some distance up the sides, maple, black birch, and other hard woods are the prevailing timber."

From the Umqui up to this chain of hills, and on the east side of the Matapedia from Fifty-Six Mile Brook to the foot of the lake the timber is mixed wood and the soil generally good.

"Along the shore of the lake, and extending inwards as you approach the upper end, fir, cedar, poplar, spruce, small juniper or tamarac, white birch, ash, and white pine are found; the ground is swampy, with low ridges of dry ground in places covered with mixed and hard wood; from the northern slope of the hills mentioned to the lake, and across the Nemtaye to the line dividing the Seigniory from the Crown Lands, the same character prevails, rendering the ground in this part of the seigniory of little value; at its upper or northern end very good land is found. My instructions not authorizing it, I did not examine the ground on the eastern side of the lake; its general appearance is rugged.

"In this section, a surface of more than one hundred square miles, (sixty-three of which are seignorial,) three-fifths are fit for cultivation: that is, twenty-four thousand in the seigniory, and fourteen thousand acres in Crown Lands."

From the Seigniory of Matapedia to that of Metis, the country is undulating, the hills seldom attain an elevation over two hundred and fifty feet above their base, with flats generally of considerable extent on top. Near and on the summits white, black and yellow birch, maple, and rowan trees prevail; on the sides the same kinds of wood with a greater mixture of fir, spruce, pine, and cedar; in the hollows and swamps, cedar and other soft woods, elm, ash, and tamarac are found but not in abundance.

"In valleys and hollows through which the streams flow, there are a number of small lakes. It is difficult to convey a general idea of their form and the appearance of the hills without inspecting a plan of the ground.

"In many places the soil is full of small angular pieces of rock, and deficient of depth, in others it is sandy: in the hollows and swamps there is a deposit of black mould from six inches to three feet in depth with clay or a hard subsoil underneath: on the higher grounds the soil is generally a yellow loam; it may be considered fully equal in quality to the greater part of the country south of the St. Lawrence, east of Quebec.

"About thirty-eight thousand acres, or rather more than three-sevenths of one hundred and thirty square miles, the extent of this section, may be considered good arable land.

"The line passes through a portion of the seigniory of Lepage-Thivierge, before reaching the River Metis; the ground in the seigniory extending ten miles back from the St. Lawrence, and in that of Metis, and the Fief of Pachot, six miles in depth, is quite as good as in the section first described.

"The extent of available ground within a width of ten miles between the Rivers Restigouche and St. Lawrence, without including that on the east side of Lake Matapedia or in the Seigniory of Metis, Lepage-Thivierge, or Fief of Pachot, may be underrated at two hundred and thirty-eight thousand acres in Crown Lands, and twenty-four thousand in Seignorial; as it is not necessary that every portion should be fit for the plough, reserves for fuel, fencing, and also building timber being required, even if this were the case.

"It may be here mentioned that a deposit of marl exists at one of the small lakes on the Nemtaye, and will in all probability be found in other places. Peat, another valuable manure, is found in different parts of the districts. Limestone is abundant at the head of Lake Matapedia and on its south-west side, and for some distance down the river * * *

"The climate of this portion of Canada does not differ materially from that of Quebec, though rather cooler in summer; intense cold is not so frequent; rainy weather or thaws of long duration do not occur, however, in winter. Snow is expected about the 22nd October, this does not remain longer than a day or two at furthest, and is followed by fine weather with one or two falls of snow, to about the 21st November, when the winter may be said to begin. The depth of snow in ordinary winters, is four feet: it has been known to reach six feet.

"Cultivated land is clear of snow about the 20th of April; ploughing commences from 1st to 8th of May Rye, wheat and peas are sown from that time to the 28th May; oats to the end of the month; barley and potatoes to near the end of June; reaping generally commences about the 25th August, and lasts to the end of September, when the potatoe crop is fit to house.

APPENDIX C.

(FRONTIER ROUTE, LINE NO. 1.)

From a Report by Mr. T. S. Rubidge, on an examination of the Country between River du Loup and Woodstock, 1860.

I have the honor to report on the character of the country and facilities for constructing a Railway from River du Loup to connect with the New Brunswick and Canada Railway, at or near Woodstock.—I wish to state that the examination was of a general character. And I beg to refer you to the accompanying map, whereon I have marked in red the route in my opinion, most eligible for preliminary survey. Although I have not personally explored the whole of the country traversed by the proposed line, more particularly the section south of the Grand Falls,—yet I have reason to believe a practicable line, nearly approximating to that indicated on the map, will be discovered, and I was sufficiently near it to enable me to speak with a degree of accuracy as to distances.

DIRECTION OF THE ROUTE RECOMMENDED FOR SURVEY.

River du Loup to Province Line, 63 miles.

Commencing at the Station, the line crosses to the east side of the Temiscouata Portage, and running towards St. Modeste, enters the valley of River Verte; thence following this Valley, it ascends continuously to the 12th mile, the summit of the dividing ridge between the waters of the St. Lawrence and the Bay of Fundy.

Again crossing the Portage the line runs nearly parallel with it to Blue River, thence assuming a direction to cross the Calaneau River near the Falls, and afterwards strikes the head waters of the River aux Perches, it descends in the valley of that stream to the Dégelé settlement on the west bank of the River Madawaska. From this point to the Province Line the route lies along the level margin of the river.

Province Line to Grand Falls, 50 miles.

Continuing down the valley of the Madawaska and crossing the river above the rapids at Little Falls, the line enters the valley of the St. John through a depression in the high ground in rear of the village of Edmundston, and it thence follows the east bank of the River St. John, crossing it a short distance above Grand Falls.

Grand Falls to Woodstock, 70 miles.

The Engineer of the New Brunswick and Canada Railway has furnished me with the following information :—" Having lately made an inspection of the country from the south bend of the Meduxnikeag River to the crossing of the Presqu'isle River, I am enabled to state that the character of the country is much the same as that portion which has been already surveyed, and I am inclined to the opinion that the road can be constructed at nearly as moderate a rate as that at which it has been already executed. There are two routes open to the line in crossing the Presqu'isle, viz : the upper route keeping to the westward of the Williamson Lake, and crossing the river near the Tracy Mills, and thence onwards to the bend of the main river,—and again the lower route taking to the eastward of the Lake, and crossing the river about one mile below the present bridge, and thence toward the main river bank. From this point to the Grand Falls along the margin of the main river the country presents a most favorable contour, the works of chief magnitude on the entire route consisting merely in bridging the Presqu'isle and Aroostook Rivers."

Woodstock to St. Andrews, 87 miles.

The line has been located to Canterbury, 22 miles : thence to St. Andrews, the railway is open for traffic.

GENERAL DESCRIPTION OF THE ROUTE—CHARACTER OF THE COUNTRY, &C.

Abstract of Distances.

River du Loup to Province Line, 53 miles, not surveyed,
Province Line " Grand Falls, 50 "   "
Grand Falls " Woodstock,   70 "   "
Woodstock   " Canterbury, 22 "   "   located and in progress.
Canterbury   " St. Andrews, 65 "   "   opened for traffic.

River du Loup "St. Andrews, 270 miles.

From River du Loup to Dégelé at the foot of Lake Temiscouata is perhaps the most difficult and expensive portion of the route, requiring very careful exploration and survey.

A whole season would be necessary to perform this service satisfactorily, as in the event of the line recommended, proving unfavorable, it would then become necessary to examine the country in the direction of the dotted line on the map.

The chief difficulty to be surmounted, is the dividing ridge or water shed between the St. Lawrence and the Bay of Fundy.

This summit elevation, 880 feet above the sea, is unavoidable; but the route by the Lakes des Roches and the St. Francis is favorable, inasmuch as it only exceeds by 100 feet the Trois Pistoles summit, the lowest yet ascertained.—From the River du Loup Station, 320 feet above the sea, the ground rises in terraces, separated by short steep slopes or rocky ridges.

These terraces are traversed by streams flowing parallel with the St. Lawrence, and are necessarily crossed nearly at right angles. It is therefore supposed that the works on this section will be of an expensive character. South of the summit to the Dégelé the country is crossed and intersected in every direction by rocky ridges or bold rugged hills, which in some instances attain an elevation of 1800 feet above the sea.

The general elevation of the ground at the base of these hills varies from 670 to 900 feet above the sea.

Owing to the broken character of the country it is supposed that a large proportion of the line will be curved, and that in extreme cases curves of half a mile radius will be required.

And long maximum gradients estimated at fifty feet per mile will be of frequent occurrence.

River du Loup is the only important stream crossed, all other streams with the exception of the Cabanceau and River Verte are crossed near their sources. The bridging will therefore be unimportant, but as a general rule the approaches will be heavy.

The total length of bridging will probably not exceed 750 feet lineal. Timber of good quality is abundant, but stone suitable for building will not be readily obtained.

The rock formation is chiefly Gneiss, Clay Slate or other similar rocks.

The soil is gravelly and frequently very rocky, but there is much excellent land on the route still ungranted.

The timber is generally Spruce, Pine, Birch, Cedar and occasionally Maple.

Settlements extend about six miles back of River du Loup, thence to the Dégelé the line runs through an unbroken forest.

The proposed route is generally within 3 miles of the new "Temiscouata Portage," therefore materials for construction or supplies for labourers will be obtained without much difficulty.

And the west shore of Lake Temiscouata from the Cabanceau to the Dégelé is partially settled; there is also a Grist and Saw Mill in this neighbourhood.

Lumbering operations are carried on to some extent on the tributaries of the St. John and Lake Temiscouata, and water power is abundant in this section of the country. From the Dégelé to Grand Falls, the country is remarkably favorable for railway purposes.

The valley of the Madawaska is generally flat or slightly undulating and its average elevation above the sea 500 feet; it is skirted on either hand by a continuous range of high steep hills which near the Province Line and in the vicinity of Edmundston approach the river.

These hills may however be avoided without difficulty, but the present highway may possibly be interfered with.—This portion of the line will be found very direct, the Grades light and the curves of large radius. Settlements occur at frequent intervals all along the west bank of the river, and towards Edmundston on the east bank also.

Thus far the settlers are chiefly French Canadians.

The village of Edmundston is situated at the junction of the Madawaska with the river St. John, and promises to become a place of some importance as a Lumbering Depôt. The river St. John is here the boundary between New Brunswick and the United States.— Both sides of the river are settled as high up as the river St. Francis, and several first class Saw Mills have recently been erected which manufacture lumber for the St. John and American markets.

From Edmundston the line will continue down the valley of the St. John, at very favorable grades, passing through a comparatively well settled, fertile, and level part of the country.

And long straight lines and curves of large radius may also be obtained here.

The banks of the St. John are alluvial, rising successively in steps towards ranges of highlands lying parallel with the river.

The rocks throughout this section of the country belong to the primitive formation. Roofing slate has been discovered near Green river.

Limestone suitable for lime has also been found.

The soil generally is a stiff clay.

The streams to be crossed are unimportant, but their valleys are sometimes very broad, necessitating heavy embankments. A great part of the route will be through cleared land. The vacant lands are usually 2 or 3 concessions back from the river.

The settlers in the Madawaska territory, which includes both sides of the river between Edmundston and Grand Falls, are Acadian French.

Near Grand Falls the country becomes broken and rocky, and is thinly settled.

A favorable site for crossing the river St. John occurs about a mile above the Falls, the banks are high and steep, and the stream narrows to a width of less than 500 feet. But much careful examination will be necessary before selecting this crossing. The bridging on this section will not, it is supposed, exceed 1000 feet lineal, including the St. John and Madawaska, the only important streams crossed. The elevation of the river in the upper basin or reach is about 420 feet above the sea.

Colbrooke, the shire town of the County of Victoria, is located on the west bank of the river, opposite the Falls; and immediately below them a suspension bridge of 190 feet span is now being constructed by the Government, the stone for the work is quarried on the spot.

Grand Falls is a formidable obstacle to lumbering operations, the river falls 74 feet over a perpendicular ledge of slate rock into a narrow gorge, nearly a mile long, descending in that distance 45 feet or 119 feet in all.

Square timber and saw logs are run over the falls, entailing a loss of 10 or 12 per cent. thereby, but all sawed lumber has to be hauled across the portage, between the upper and lower basins, as also all supplies going up the river.

In New Brunswick lumbering operations have gradually receded, and now lie chiefly on the waters of the upper St. John. The proposed Railway would certainly promote the settlement of this most valuable timber region. It would also develop the manufactured lumber trade by affording facilities for obtaining supplies and for transportation to market, either at St. Andrews, Quebec or River du Loup. It would create in the interior of New Brunswick and the State of Maine a market for Canadian provisions, and thus open up a new trade with Montreal and the cities farther west. Saw Mills for manufacturing timber would be erected on the tributaries of the St. John, and eventually almost all the timber on the river would be converted into Deals, Clapboards, Shingles and similar short lumber. The lumbering establishments on the upper St. John and Lake Temiscouata require very large supplies of Flour and Pork which (with the exception of a small quantity obtained direct from Quebec, by the Colonization road and Temiscouata Portage) are usually sent by Steamboat or Railway to Woodstock, and are thence forwarded up the river in flat bottomed boats towed by horses. At present the supplies and merchandise forwarded up the river is stated to be equal in bulk to 80,000 Bbls. (Flour.)

Distributed as follows : { 30,000 Barrels to Woodstock and vicinity.
30,000 " Tobique and Aroostook.
20,000 " Grand Falls and upwards.

From Grand Falls to Woodstock is said to be one of the most productive agricultural districts in New Brunswick, but the country appears rough and unfavorable for Railway construction, being intersected by very deep valleys and ravines, through which flow streams leading into the river St. John. The surveys of the New Brunswick and Canada Railway extend only to the Little Presqu'isle River, 10 miles north of Woodstock, and it is reported "from this point forward the surface of the country is comparatively level."—The vacant lands in this section of the country lie beyond the settlements on the eastern bank of the St. John.—The population of the River St. John above Woodstock, including the Aroostook country, is estimated at 10,000. The inhabitants of the county of Aroostook, in the State of Maine, are much interested in the proposed Railway—Their most important lumber streams flow into the St. John, and many of the roads leading from the interior of the country connect with the " Great Roads" of New Brunswick.—This portion of the state is rapidly becoming settled by a large farming population, it is also a most valuable timber region abounding in water power.—From the great quantity of lumber manufactured for the American Market, as well as the supplies required for lumbering operations, the Aroostook country must eventually prove a most important feeder for the Railway.—The amount of lumber, &c., produced and annually sent down the river is stated to be nearly as follows, viz :

Square timber from above Grand Falls.................... 4,000,000 feet.
"   "   "   below   "   .... 3,000,000 "
Sawed lumber from Aroostook, { Shingles......... ....20,000,000 No.
Clapboards .... 1,500,000 "
Boards........................... 750,000 "
Oats....................... 10,000 bushels.
Potatoes........................ 5,000 "
Buckwheat Meal................ 60 tons.
Oat............................ 30 "

Woodstock, the shire town of the county of Carlton, is situated on the west bank of the St. John, at the mouth of the Meduxnikeag River, and at the extremity of a " Great Road" to Houlton, Maine, on which there is much traffic. Both towns are of considerable importance as being the centre of a large agricultural population. Extensive Ironworks were formerly in operation near Woodstock, copper has also been discovered in the neighbourhood. From Woodstock to Canterbury, the present terminus of the New Brunswick and Canada Railway, the distance will be either 22 or 25 miles, dependent on the route adopted, relative to this section, I extract the following information from the report of the Engineer and Manager. The location from Eel River to Woodstock is not yet decided upon, consequently no work has been commenced north of the former place. Two lines have been surveyed, one running direct to Woodstock the other to the Houlton road, which it crosses nearly midway between Woodstock and Houlton. From Eel River direct to Woodstock, involving at the commencement grades of 50 feet for 2 miles from Eel River, at which point the summit is attained, and from which there is a descent all the way to Woodstock ; some heavy work has to be encountered in crossing the wide creeks, which cannot possibly be avoided or materially reduced by any diversion of the line : nevertheless the quantities of excavation are comparatively light, and the general direction good ; through 16 miles of Forest, and 6 miles cleared land, there is no curvature of less radius than 1910 feet, and only three of these to Woodstock. The grades may also be considered as favorable, the maximum being 62 feet to the mile for one mile, and in the direction of the down traffic. This is, without exception, the most practicable route from Eel River to Woodstock. The comparative estimates however exhibit the cost of construction as £37,527 in excess of those of the upper routes by the Houlton Road. We may also mention in connection with this route, that its extension beyond Woodstock by way of the eastern branch of Lanis Creek, is also the most favorable and practicable egress that can be found over such a very rough country as presents itself in that vicinity : for 10 miles northward, 65 feet grades are absolutely necessary to reach the summit level, the only redeeming qualification, being that the declivity is to the south towards St. Andrews, and is therefore favorable to the down traffic.

The work on the first 10 miles section from Canterbury is of the heaviest character.
From Canterbury to St. Andrews is 65 miles.
The road is said to be completed and in good running order.
The number of way stations including Canterbury is 12.
The Guage is 5ft. 6in., uniform with the European and North American Railway (St. John and Shediac.)—I was unable to obtain reliable information as to grades, curves or permanent way.

Embankments are 15ft. wide at formation level, slopes 1½ to 1.
Earth Cuttings " 30 " " " "
Rock " " 24 " " " vertical.

Bridge abutments of Ashlar Coursed, or in coursed Rubble.
" Superstruction of Timber.
Culverts are of Cedar Timber or dry rubble masonry coursed.

The Company has a Grant from the Government of all vacant lands within a distance of 5 miles on either side of the Railway. A large proportion of these lands are represented as being very valuable as well for agricultural as for lumbering purposes. It is stated that the harbour of St. Andrews is occasionally frozen, also that the depth of water at the entrance is insufficient. The first statement is incorrect. But with reference to the depth of water it is stated in the Report of the Board of Works for 1858, that 40,000 c. yds of dredging might perhaps be sufficient to make the entrance of the Harbour available for a depth of 8 feet at lowest spring tides, this would enable a vessel drawing 20 feet to come into the Harbour at half tide. Spring tides rise from 24 to 26 feet, and neaps from 20 to 22. Chamcook Harbour about 4 miles N. E. of St. Andrews, appears well adapted for Ocean Steamers. The Railway is said to skirt the shore of this Harbour.

APPENDIX D.

(FRONTIER ROUTE, LINE No. 2.)

Correspondence in reference to the extension of the St Andrews and Woodstock (the New Brunswick and Canada) Railway to River du Loup.

St. Andrews, 5th September, 1864.

DEAR SIR,

On my arrival in Town on Saturday evening last, Mr. Osburn placed in my hands your letter to him of the 20th ult., in which you express a desire to be furnished with a copy of my Report of a Survey conducted by me during the Winter of 1861, for the extension of the St. Andrews Railroad to the Canadian Frontier.

I have now great pleasure in presenting you with copies of Reports I then made, and gladly avail myself of a brief sojourn at home, to put you in immediate possession of any useful information they may contain.

Your very truly,
WALTER M. BUCK.

SANDFORD FLEMING, ESQ.,
   Civil Engineer, &c., &c.,
      Tobique.

St. Andrews, N. B., 3rd February, 1862.

HENRY OSBURN, ESQ., Manager.
Dear Sir,

I beg to submit the following Report upon the Preliminary Survey recently made in two sections, viz: from the south branch of the Meduxnikeag river (at which place the former Richmond-Corner and Hillman-Valley locations terminated) to the St. John River at Wilson's, and from the Grand Falls southward to the Tobique river at Hutchinson's.

This survey was commenced on October 15, 1861, and was continued to the 7th of January, 1862, but was not completed at this period; the section of country between the river St. John, at the proposed crossing place at Wilson's by the Hardwood Creek, and thence by the Valley of the Menguart river, and over the summit ridge, which divides the head waters of the latter from that of the Trout brook and Otellock river, to the Tobique river being left untouched; as also the section of country north of the Grand Falls to the Canadian Frontier.

The greater portion of this proposed route from the river St. John, has been traced on foot through the Woods, in company with a small party necessarily organized for such an expedition, amongst whom were men whose knowledge of the localities, obtained from lumbering operations, justified their engagement, whilst others were employed for the purpose of sacking or carrying the camp equipage and provisions. The time occupied in making this exploration to within a few miles of the Canadian Frontier, from leaving St. Andrews, was forty days, and you will observe from the copious notes taken during this period, that the examination was carefully made, although under many difficulties, arising from the continued inclemency of the weather. The surveying party on the section from Richmond forward, under the direction of Mr. Chas. Haslett, received instructions to pursue a route that was considered to be the most eligible and practicable in the direction of the river St. John, this portion of the country having been better known from previous travelling.

The other party, under the direction of Mr. John Otty, were sent torward to the Grand Falls, and received instructions to commence the survey at that place, and on the west side of the river, working southward, until it should become known from a reconnaissance on the east side of the river, through the interior of the country, whether a line of road was practicable or not from the Tobique river to the Grand Falls; the examination having established the affirmative, the surveying party were ordered to abandon their work on the west side of the river, with which they were progressing most favorably, and to commence fresh operations on the east side, near the head of the Mooney brook, a tributary to the Big Salmon river.

The Munguart river and Trout brook district was also examined : the Valleys of these waters are intercepted by a summit ridge, which will require more precise instrumental exploration, than could otherwise be made, to ascertain the maximum grades that will have to be adopted ; on the other two sections the maximum grade is but 53 feet per mile. It was intended to have contour levels taken over this portion of the route, and also all other levels properly connected and reduced from one Datum, but unfortunately the surveying parties had to abandon all further operations on account of severe snow storms and other causes. It would however take but a short time to connect the whole work by these levels at an early and more favorable period, the expense incurred would be but trifling in comparison with the great importance of having continuous levels and known relative elevations.

The section of country between the Grand Falls and the Canadian Boundary was next explored, and proved the most favorable for Railway construction. The general proposed direction will be by the Valley of the Dead-brook, and Second Beaver-brook, crossing the Grand river on its marginal flats, thence by the Sigus-lake and branch across the Sigas-river, and stretching almost directly across to the forks of the Quisibis river; thence across the Green River to the front of the Green mountain, and approaching the main river at St. Bazil, which will be the nearest touching point ; and then along a table-land at the foot of the Green river ridges to the Iroquois river, and up the Valley of this river to the Canadian boundary, where Mr. Rubidge, the Engineer in charge of the Canadian Survey, terminated his explorations, having pronounced the former proposed route to the westward of the Temiscouata lake, on instrumental examination, to be entirely impracticable.

Your attention is particularly requested to the accompanying map, shewing the line of the Halifax and Quebec Railway and its connections, &c. ; it has been taken from a published pamphlet "On the political and economical importance of completing the line of railway from Halifax to Quebec," by Joseph Nelson. You will observe that the yellow tinted line, being the proposed central line for the Intercolonial railway, is traced to the westward of the Temiscouata lake, evidently shewing that at the time the map was prepared and the proposed route marked thereon, nothing was then known of its actual practicability ; the same may be said of that portion also which is lined between the Tobique river and the Dégelé, at the foot of the Temiscouata lake. During the recent exploration, Green Mountain, which is said to be upwards of one thousand feet above the St. John river, was ascended to its snow-clad top, and the view of the country to the eastward and northward was sufficient to impress me with the impracticability of extending a road on that side of the mountain, through such a mountainous region ; when I say impracticable, I mean by it a most unjustifiable expenditure in construction.

Herewith is also furnished a profile of 17 miles of the survey between Grand Falls and Tobique river, likewise an estimate of the cost of construction of—

50 miles of the proposed route amounting to £295,000 cy.
That of the first 30 miles, averaging per mile 5,440 Stg.
And that of the other 20 miles " 3,643 "

These estimates may be received as full and ample for the respective sections only, and I trust that so far as this winter survey has been extended, the result will be considered satisfactory.

WALTER M. BUCK,
Engineer in charge of Survey.

St. Andrews, N. B., 8th March, 1862.

HENRY MAUDSLAY, ESQ.,
of London,
Board Director N. B. and C. R. R.

DEAR SIR,
In accordance with your request I beg to submit the following Report as supplementary to that of 3rd February last.

The site intended for the Station buildings at the Richmond terminus (so called) is at McGeorge's, on the Hillman Valley; the grounds will be level for 1800 feet and can be graded on embankment to any extent in width that may hereafter be required; this portion was selected, as at first proposed, in consequence of a heavy ascending grade of 56 feet per mile being required to reach the summit at the Houlton and Woodstock road in a deep cutting and would not be suitable for the approach to the station.

The descent from the summit to the Valley of the Meduxnikeag river is made by adopting steep gradients, one of 60 feet per mile being employed for a short distance.

From the point of intersection with the high road the distance to the Woodstock is reckoned as 7 miles, and to Houlton 5 miles; Houlton is situated about 3 miles within the boundary line.

The preliminary survey recently made for the extension of the line northward, was carried to within 3 miles of the St. John river, at Wilson's, opposite the Hardwood Creek, at which place, the crossing will necessarily be on a high level of about 100 feet above water surface, the width of the river being fully 800 feet. The partial location made was twenty-seven and a half mile through a thickly wooded country, and in order to obtain correctly the positions and elevations of points through which it was desirable to pass, the public and bye roads were traversed, and levels taken; forty-three miles of this work has been accomplished in addition to the other work, and from which a topographical plan of this portion of the country can be made whenever required.

At the south branch of the Meduxnikeag river, which has its rise in the State of Maine, and joins the St. John water at Woodstock, the line crosses above the Falls, and at a level of 55 feet above water surface. The fall of the river to Woodstock is about 215 feet in a distance of 8 miles, or thereabouts, so that a branch line into Woodstock along the Valley of this river would be perfectly practicable; the total distance to this point from St. Andrews is 96 miles.

The north branch of the Meduxnikeag river is next crossed at the 98th mile, with an ascending grade, adout 35 feet above water level; the crossing is almost on the square and a little below the third falls, and over solid rock; both sides may be considered as natural formed abutments for bridging.

The location from Fulcan's on the 92nd mile and for about three miles forward, must of necessity approach and run parallel to the boundary line within a mile distance, and at the crossing of the Meduxnikeag south branch within one and three quarter mile. From the north branch the line takes an easterly course and crosses the little Presqu'isle river at the 106th mile, in the Williamstown village, this stream flows from the Williamstown lake to the St. John river, about 6 miles apart. The lake is a fine sheet of water two miles in length, and one mile in width. The village of Williamstown is about 14 miles from Woodstock, and within 5 miles of the boundary line, the river at this place affords excellent water power for Saw mills, and the village would, no doubt, become a thriving place when accessible by railway.

From this point forward the location takes a northerly course with uniform grades, to within 2 miles of the big Presqu'isle river on the 112th mile. This river which has its source in the State of Maine is crossed on the level 75 feet above water surface: it is approached from the south with a 49 feet grade, and from the north with a 53 feet grade; the point of crossing is within 2 miles of the St. John river, and six miles of the boundary line, and pursues a northerly course to the St. John river, at Wilson's, in Upper Wicklow, opposite the Hardwood Creek.

The location was not completed to this point, but as the public roads were traversed, and an exploration made through the woods, it was concluded that the character of the country did not vary much, and the estimates were framed upon the same average quantities per mile.

From Fulcan's on the 92nd mile to the St. John river on 120th mile, the quickest curvature necessarily employed is 3° or 1910 feet radius, and this between the branches of the Meduxnikeag river, and to within a mile of the Florenceville road (14 miles beyond the Meduxnikeag) the location chiefly consists of tangents, no quicker curvature being required than one mile radius; and from Florenceville to the St. John river, the location is also principally on tangents, the sharpest curvature being half a mile radius.

Three fourths of this section has been partially located and presents 20 miles of straight line, 5 miles of 1° curvature or 5730 feet radius, and 5 miles of 2°, 2° 30′ and 3° curves, the radii being 2865 ft., 2292 ft., and 1910 ft.,; the maximum gradient is 53 feet per mile.

The quantities estimated on this section are for Earthwork 26,000 cubic yards, and for rock 1666 cubic yards per mile. The total estimated cost of construction including masonry, bridging, ballasting, superstructure and station buildings, &c., will average £5,500 Stg. per mile.

The banks and bed of the St. John river, at the proposed crossing consisting of rock formation, and the narrowest place as well, it is admirably adapted for bridging, more especially as there is a fine granite quarry in the immediate vicinity. The approaches on either side of the river will involve heavy embankments, but the grades will be favorable.

The next portion of country between the St. John and Tobique rivers, through which the line would traverse, has not been surveyed, and but partially explored; this length of line will be about 26 miles. After leaving the Hardwood creek which heads in the Moose Mountain range, it follows in a northerly direction the valley of the Munguart river, and crosses northerly the dividing ridge between the head waters of the tributaries to the St. John and Tobique rivers; it then continues by the head of Trout brook and takes the valley of the Otelloch river for some distance, then diverges across to the Tobique river below the mouth of the Otella river. No levels have been run over this district, consequently no profile has been furnished, and the summit level has not been ascertained.

On reference to the Map it appears that the proposed route for the central line is laid down to cross the Tobique river, seven miles upstream near to the Wapskehegan river, and the Major Robinson central route crosses as far up as the Gulquac river; both these lines pass through a more difficult country than that in the neighbourhood of the Munguart, as the eminences in the range of the Tobique mountains increase in altitude as you ascend the river up to the Blue Mountain, about 50 miles from the mouth. The country between St. John and Tobique rivers is thickly wooded; spruce and birch being the predominant growth; the land is not settled upon within the banks of the river, but it is pronounced to be of good quality.

The survey of the section between the Grand Falls and the Tobique river, the party working southwards, commenced on the 28th October last, the distance being about 20 miles through an unbroken wilderness. A line was first started two miles to the eastward of the Grand Falls, and run along a valley to the Salmon river, in the direction of the Little Salmon: this was taken as the shortest line, but as the first stream could not be crossed to advantage without adopting a 70 feet grade to descend from the summit within two miles, which was considered objectionable, although not strictly so upon a trial-survey, the line was abandoned, and a position taken up three miles still further to the eastward of the Falls near the head of the Mooney-brook, being a much lower level than at first chosen. The descent of the brook is made with a 53 feet grade for two and a half miles to its mouth, the Salmon river being crossed at a level of 22 feet above water, with the same grade continued to the end of the third mile.

A succession of uniform grades with light work is then continued to the crossing of the Little Salmon at the forks on the 6th mile and from this point an ascent is made up the Valley of the stream to its head, and that of Little river (a small stream flowing to the St. John) and to the summit level on the 16th mile; the total rise being 354 feet in nine miles, or an average grade of 39 feet per mile, but on account of a level interval occurring, a grade of 53 feet per mile has to be introduced for nearly half the distance.

Little Salmon river is a very tortuous stream, and it will be necessary for the line to cross it frequently, unless bridging can be dispensed with by making diversions; it can be spanned by a 30 feet girder bridge at any place.

Some rather abrupt land occurs near to the summit, but it is the only heavy work (by comparison) on the whole of this length, viz: an embankment containing 50,000 cubic yards, and a cutting 2000 feet in length, with a maximum depth of 25 feet.

After passing over this summit the line falls into the Valley of the Bear-brook on the 17th mile, and within about 3 miles of the Tobique river at Hutchinson's, at which place the river is probably 400 feet wide.

The quantities estimated are, for earthwork, 18,350 cubic yards, and for rock 1,150 cubic yards per mile, the estimated cost per mile for all materials as on the Richmond section is about £3,650 Stg.

It is to be regretted that this survey was commenced at such a late season of the year, the snow being at the deepest, and the days at their shortest; had it been taken in hand during the summer or the fall of the year, double the amount of work could have been performed to much better advantage, and provisions would have been at lower prices; however as it was a necessity at the time instructions were first received, it can only be said that all that human effort could accomplish in the woods at such a period, was done.

In addition to the foregoing I beg to refer you to my Report, dated 3rd February last, addressed to the Manager, and forwarded by him to your Board of Directors.

WALTER M. BUCK,
Engineer in charge of Survey.

APPENDIX E.

(CENTRAL ROUTE, LINE NO. 8.)

Report on Exploration from the Village of Boiestown across the Tobique Highlands.

SANDFORD FLEMING, ESQ.,
        Chief Engineer,
                Intercolonial Railway.

DEAR SIR,

In accordance with instructions, verbal and written, received from you in March last, I proceeded to make an exploration of the country from the village of Boiestown, northward to the sources of the Dungarvon, Rocky Brook and Gulquac rivers, and now beg leave to hand you the following remarks :

Having placed an Aneroid Barometer in the hands of a careful party at Boiestown, with instructions to note its changes at certain periods of the day, and to record name on a table previously prepared by myself; I started for the point previously arranged, (viz.) the boundary line between the counties of York and Northumberland, and immediately west of the Upper Falls of the main Dungarvon, commenced operations by running a series of lines diverging from this point in order to ascertain the main features of the country ; I found however that these lines so frequently carried me over the tops of high mountains, that it would be necessary to adopt a different system of working, and confine my explorations to the several streams, which in this part of the country cannot be said to run through valleys, but merely Gorges varying in their breadth from the simple width of the river to perhaps a quarter of a mile and bounded on both sides with high land broken only by the defiles of the few mountain streams that feed the main rivers.

Having decided on the above line of operations I first traced the main Dungarvon from a point about three miles below the " Upper Falls" to its sources, the most northerly of which I found to be at an elevation of 1215 feet above Boiestown ; I then followed a branch of this stream running in a northwest course from the vicinity of the " Upper Falls," and found it to head in still water to the west of the county line before mentioned and continuing on passed over the dividing ridge between the Dungarvon river and the Rocky Brook, at an elevation of about 930 feet ; from this point I followed two valleys or gorges running in different directions to the Rocky Brook around a high hill as you take notice at Obs. No. 33 ; the Rocky Brook on the west side of this hill passes between very precipitous rocky banks, which would render the building of a railway at this point an expensive matter, this can however be avoided by following the two valleys mentioned ; continuing on up the Rocky Brook I first explored the right hand branch which, after passing between very precipitous rocky banks, and over these Falls, takes its rise in a large lake at an elevation of 1118 feet, quite surrounded by high hills, through which I could not see any depression, at least in the direction that I wished ; returning to the Forks, followed up the left hand branch and found it to head in a Lake at an elevation of about 950 feet, passed on over a dividing ridge of about a quarter of a mile in length, and at a height of 965 feet, and entered upon the head waters of a branch of the Clearwater Brook, followed it for several miles through Lakes, Streams and Beaver Dams, &c., till it reached the Main Stream, thence up this stream to its source which I found to be in a Swamp or Barren at a height of 1513 feet, this being the summit level between the Clearwater Brook and the Gulquac River.

On the annexed sketch I have put a number of heights with the number of the observation above it for the guidance of any party that may be sent out to carry on the detail survey; all my observations are marked on Trees with red chalk and numbered consecutively, as also all the lines run are numbered as shown in the sketch.

Owing to the winter being so far advanced before I started out on this survey, I was obliged to move with great rapidity from one part to another, as I found the rivers breaking up very fast and the danger of freshets setting in was every day increasing, this of course prevented me exploring the country as far or as minutely as I had at first intended; and added to this rapid breaking up of the streams, I was still further impeded by the continuance for a whole week of a snow storm just at the time that I was in the region of the head waters of the Gulquac and Clearwater; this rendered any attempt at a topographical delineation of the country impossible. I have, however, laid down some of the features of the country thereabouts as far as was possible from lines run under the circumstances, and have also sketched on in blue ink the most probable route for a Railway Line through this section of country, which, so far as my explorations extended, shew it to be quite practicable from the Miramichi side, but owing to the sudden breaking up of the streams, I did not deem it prudent to venture further into the country, consequently I returned by the shortest route (viz.) the Wapskehegan river, down which we were obliged to travel on rafts or catamarans; this of course prevents me giving you any correct report of the country along the Gulquac, but from what little I saw of it and the height of its head above its junction with the Tobique which cannot be more than 550 feet in a distance of about fifteen miles, places this route quite within the range of practicability.

Owing to the depth of snow on the ground, I had not an opportunity of judging of the soil for agricultural purposes, but from the timber found on the high lands (Birch and Maple), I should deem it to be of a character suitable for such uses; but the lower levels and barrens were generally covered with Cedar, Spruce and Hacmatack; the most of the country travelled over by me will yield good building material for the ordinary structures used on a Railway.

In conclusion I may add that the general features of the country are favorable for the construction of a Railway, as the banks of the streams in most cases recede from the water at a uniform rate of inclination.

I am,
Yours truly,
W. H. TREMAINE.

Halifax, May, 1864.

APPENDIX F.

REMARKS on the shortest lines of Communication, between America and Europe, in connection with the contemplated Intercolonial Railway.

In the Northern United States many leading men who take a prominent part in directing the great works of intercommunication of the country, have long aimed at an extension of their Railway System to some extreme eastern Port on the Continent; their object being to shorten the Ocean passage and the time of transit, between the great commercial centres of the Old and New Worlds.

A plan was propounded in 1850 by which it was proposed to connect the cities of New York and Boston with Halifax, by a Railway stretching across the State of Maine, the Provinces of New Brunswick and Nova Scotia.

The originators and promoters of this plan correctly assumed, that the necessities of trade, would sooner or later require the adoption of the shortest possible sea voyage between the two Continents.

This scheme appears to have found no little favor in New Brunswick and Nova Scotia.

The line of Railway then projected was designated "The European and North American Railway," hence the name of that important section of it, constructed and in operation, between St. John, New Brunswick, and the Isthmus which connects that Province with Nova Scotia.

The whole scheme as originally proposed has, ever since its projection, been kept prominently in view; and there only now remains to complete it, the link between Moncton and Truro (common to the Intercolonial Railway) and that other link between St. John and Bangor, so warmly advocated at the present time in the States of Maine and Massachusetts. The whole project has still many advocates in both the Provinces referred to.

These Railway links completed, the city of Halifax would be connected with the whole of the United States, and the Ocean passage between the Railway systems of Europe and America would be reduced to the distances between Halifax on the one side, and Galway, or some other Port on the west coast of Ireland, on the other.

It is a question, however, if Halifax would permanently remain the Entrepot for Ocean Steamers. The same consideratious which so strongly influenced the originators of "The European and North American Railway," and which still so powerfully weigh with its promoters, would induce them or their successors to look for a point of embarkation still nearer Europe than Halifax.

Halifax might then have to give way to the most easterly Harbour in Nova Scotia; and should the bridging of the Gut of Canso not defy engineering skill and financial ability, the great European Terminus of all the Railways on this Continent might yet be situated on the Island of Cape Breton.

There are two good Harbours on the easterly coast of Cape Breton, the one at Sydney where the best of coal abounds, and the other, the Old French Harbour of Louisburg where similar advantages may obtain. Sydney and Louisburg are respectively about 100 and 180 miles nearer Europe than Halifax, and although it is said they are not open ports all the year round, yet they are undoubtedly open during the great travelling season, and whilst open, being so much nearer Europe than Halifax they would then without question be preferred.

These considerations very naturally lead to reflections on the whole subject of Transatlantic communications, and the important question presents itself: what route may ultimately be found the *very speediest* between the Old world and the New?

Newfoundland, a large Island off the main land of North America, and Ireland an Island off the European coast, resemble each other in being similar outlying portions of the Continents to which they respectively belong. Possibly they may have a more important similarity and relationship, through the remarkable geographical position which they hold, the one to the other, and to the great centres of population and commerce in Europe and America.

A glance at the chart of the Atlantic will shew that between Ireland and Newfoundland the Ocean can be spanned by the shortest line.

Ireland is separated from England and Scotland by the Irish Channel; Newfoundland is separated from this continent by the Gulph of St. Lawrence. Already railways have reached the western coast of Ireland and brought it within sixteen hours of the British capital. Were it possible to introduce the Locomotive into Newfoundland, and establish steam communications between it and the cities of America, a route would be created from Continent to Continent having the Ocean passage reduced to a minimum.

This route would not be open for traffic throughout the whole year; during certain months, the direct course of steamers would be so impeded by floating ice, that it could not with certainty or safety be traversed. It therefore remains to be seen whether the route has sufficient advantages whilst open, to recommend its establishment and use, during probably not more than seven months of the year.

In this respect the Newfoundland route must be viewed precisely in the same light as many other lines of traffic on this Continent, and possibly it may be found of equal importance. Of these works may be mentioned the Canals of Canada and the United States, which, although closed to traffic during winter, have justified the expenditure of enormous sums of money in their original construction, and in repeated enlargements and extensions.

Having alluded to the great objection to a route across Newfoundland, we may now proceed to enquire into its merits.

The track of Steamers from the British coast to New York, and to all points north of New York, passes Ireland and Newfoundland, either to the north or to the south; the most usual course, however, is to the south of both Islands. Vessels bound westerly, make for Cape Race on the south-easterly coast of Newfoundland; whilst those bound easterly, make Cape Clear on the south-westerly angle of Ireland. Not far from Cape Race is the Harbour of St. Johns, and near Cape Clear is the Harbour of Valentia; the one is the most easterly Port of America, the other is the most westerly Port of Europe. They are distant from each other about 1640 miles.

The Irish Railways are not yet extended to Valentia, but they have reached Killarney, within about 30 miles of it.

From St. Johns across Newfoundland to the Gulph of St. Lawrence the distance is about 250 miles. On the St. Lawrence coast of the Island, the Chart shews two Harbours, either of which may be found available as points of transhipment; the one St. Georges Bay, the other, Port au Port; they are situated near each other, and both are equally in a direct line from St. Johns westerly to the main land.

On the westerly shore of the Gulph we find at the entrance to the Baie des Chaleurs, the Harbour of Shippigan, mentioned in the body of the report on the surveys of the Intercolonial Railway.

From St George's Bay to Shippigan, the distance is from 240 to 250 miles. Shippigan may be connected by means of the contemplated Intercolonial Railway with Canada and the United States.

Although a very little only is known of the physical features of Newfoundland, from that little we are justified in assuming that the construction of a Railway across it from east to west is not impracticable.

Perhaps the only white man who has travelled entirely through the interior in the general direction of the projected Railway route is Mr. W. E. Cormack.

This gentleman travelled across the country many years ago, from Trinity Bay on the east, to St. George's Bay on the west. He left the eastern coast about the beginning of September, and reached St. George's Harbour on the 2nd of November.

From Mr. Cormack's account of his journey, it would appear that although a belt along the coast is hilly and broken, much of the interior is comparatively level, consisting of a series of vast savannas.*

---

* The features of the country assume an air of expanse and importance different from heretofore. The trees become larger, and stand apart and we entered upon spacious tracts of rocky ground entirely

It is more than probable that the interior may be reached by some of the Rivers or numerous Inlets, which on the map seem to pierce the mountainous belt extending along the margin of the Island.

The line of Steam communication from Great Britain across Ireland and Newfoundland, and by the contemplated Intercolonial Railway to the Interior of North America, possesses some important recommendations as will presently be seen. It will however first be necessary to allude to the question of speed.

At the present time Ocean Steamers generally carry both freight and passengers, and in this respect they are like what are termed "mixed trains" on Railways. These mixed trains are employed to serve localities where there is not sufficient passenger and freight traffic to justify the running of separate trains.

On Railways doing a large business, the traffic is properly classified; fast trains are run to carry passengers and mails only, whilst slow trains are used to convey heavy freight. A similar classification of Ocean traffic may be suggested. Freight will naturally go by the cheapest mode of conveyance, while Passengers and Mails will seek the speediest.

It is well known that the shape of a Steamship, other things being equal, governs her speed. The shape again depends on the load she may be constructed to carry: if the Ship is required only for Mails and Passengers and such voyages as need but a small quantity of fuel, she may be constructed on a model both sharp and light, and thus be capable of running more rapidly than if built to carry heavy and bulky loads. A Steamship for heavy loads may be compared to a dray horse, whilst one made specially for passengers and rapid transit may resemble a race horse, and like the latter the less weight carried the more speed will be made.

---

clear of wood. Every thing indicated our approaching to the verge of a country different from that we had passed over.

On looking towards the sea coast, the scene was magnificent. We discovered that, under the cover of the forest, we had been uniformly ascending ever since we left the salt water at Random Bar, and then soon arrived at the summit of what we saw to be a great mountain ridge, that seems to serve as a barrier between the sea and the interior. The black dense forest through which we had pilgrimaged presented a novel picture, appearing spotted with bright yellow marshes, and a few glassy lakes in its bosom, some of which we had passed close by without seeing them.

In the westward, to our inexpressible delight, the interior broke in sublimity before us. What a contrast did this present to the conjectures entertained of Newfoundland! The hitherto mysterious interior lay before us—a boundless scene—emerald surface—a vast basin. The eye strides again and again over a succession of northerly and southerly ranges of green plains marbled with woods and lakes of every form and extent.

The great external features of the eastern portion of the main body of the Island are seen from these commanding heights. Overland communication between the bays of the east, north, and south coasts, it appears, might be easily established. The chief obstacles to overcome, as far as regards the mere way, seem to lie in crossing the mountain belt of twenty or forty miles wide on which we stood, in order to reach the open low interior. The nucleus of this belt is exhibited in the form of a semi circular chain of insulated masses and round backed granitic hills generally lying N. E. and S. W. of each other in the rear of Bonavista, Trinity, Placentia, and Fortune Bays. To the southward of us in the direction of Piper's Hole in Placentia Bay, one of these conical hills, very conspicuous, I named "Mount Clarence" in honor of His Royal Highness, who, when in the navy, had been in Placentia Bay. Our view extended more than 40 miles in all directions. No high land, it has been already observed, bounded the low interior in the west.

September 11.—We descended into the bosom of the interior. The plains which shone so brilliantly are steppes or savannas, composed of fine black compact peat mould, formed by the growth and decay of mosses. They are in the form of extensive, gently undulating beds, stretching northward and southward, with running waters and lakes skirted with woods lying between them. Their yellow green surfaces are sometimes uninterrupted by either tree, shrub, rock or any irregularity, for more than ten miles. They are chequered every where upon the surface by deep beaten deer paths, and are in reality magnificent natural deer parks, adorned with wood and water.

Our progress over the savannah country was attended with great labour and consequently slow, being at the rate of from five to seven miles a day to the westward, while the distance walked was equivalent to three or four times as much.

Always inclining our course to the westward, we traversed in every direction, partly from choice in order to view and examine the country, and partly from the necessity to get round the extremities of lakes and woods, and to look for game for subsistence. We were nearly a month in passing over one savanna after another. In the interval there are several low granitic beds, stretching as the savannas northerly and southerly."

Narrative of a journey across the Island of Newfoundland, by W. E. Cormack.

If these views are correct, it is clear that the speed of Ocean Steamships might be considerably increased when constructed for a special purpose. The distance between St. Johns, (Newfoundland) and Valentia is not much more than half the distance between Liverpool and New York; and hence about half the quantity of Coal and Supplies would be required for the Passage, between the former points.

It is quite obvious therefore that a Steamship constructed specially to run between St. Johns and Valentia, and for the purpose of carrying only Passengers and Mails, with such light Express matter as usually goes by passenger trains, would attain a much higher rate of speed than existing Ocean steamers.

A rate of 16½ miles per hour is thought to be quite possible: the distance between Valentine and St. Johns is 1640 miles. At this assumed rate therefore the Ocean passage might be accomplished in 100 hours.

With regard to the speed on land, it appears from Bradshaw's Railway Guide, that the Irish mails are regularly carried between London and Holyhead at the rate of 40 miles an hour including stoppages, that the Irish Channel is crossed at the rate of 16 miles an hour, including the time required for transhipment at Holyhead and Kingstown, and that the mails reach Queenstown some 16 hours after they leave London. Valentia is very little further from Dublin than Queenstown, and on the completion of a Railway to Valentia, there is nothing to prevent it being reached from London in the same time now occupied in carrying the mails to Queenstown.

Galway has been mentioned as a proper point to connect with Ocean Steamers, it is fully an hour nearer London than Valentia, but probably three hours (in time) further from America.

Although 40 miles an hour is a common rate of speed on the Railways in England, it is not usual to run so rapidly on this side of the Atlantic.

On the leading passenger Routes in the United States, 30 miles an hour including stoppages is attained, although a rate of 25 miles an hour is more commonly adopted.

On lines frequently obstructed by snow drifts, it is not easy to maintain in Winter a rapid rate of transit, but in Summer with the rail track and rolling stock in a fair condition of repair, there is no difficulty in running at the rate of 30 miles an hour with passenger trains : and therefore this rate of speed, may reasonably be assumed as that at which the mails might be carried overland, to various points hereafter referred to on this Continent.

Having fixed upon a practicable rate of speed by land and water, the time necessary for the conveyance of the Mails from London to New York, by the projected route, may now be ascertained :

From London to Valentia at present rate of speed in England ...... 16 hours
" Valentia to St. Johns, 1640 miles at 16½ miles per hour .... 100 "
" St. Johns to St Georges, 250 miles at 30 miles per hour ...... 8½ "
" St. Georges to Shippigan, 250 miles at 16½ miles per hour.... 15½ "
" Shippigan to New York, 906 miles at 30 miles per hour....... 31 "

Total.. 171 hours.

It is thus apparent, that without assuming a rate of speed at all extraordinary, it would be possible to carry the Mails from London to New York in 171 hours, or 7¼ days, by the route passing over Ireland, Newfoundland, and by the proposed Intercolonial Railway from Shippigan.

In order to compare the route referred to with existing lines, the results of the past year may now be presented.

PASSAGES BETWEEN LIVERPOOL AND NEW YORK

Name of Steamship Line.	West'n Pas.	East'n. Pas.	Mean.
Inman Line.—Average of 52 Eastern and	d. h. m	d. h. m.	d. h.
52 Western Passages................	13 19 11	12 18 54	13 7
Shortest passages.........	11 5 0	10 5 0	10 17
Cunard Line.—Average of 27 Eastern			
and 25 Western passages.............	11 12 46	10 11 42	11 0
Shortest passages.......	9 17 0	9 3 0	9 10

PASSAGES BETWEEN SOUTHAMPTON AND NEW YORK.

Name of Steamship Line.	West'n Pas	East'n Pass.	Mean.
	d. h. m.	d. h. m.	d. h.
Hamburg Line.—Average of 23 Western and 25 Eastern passages...............	13 11 46	12 15 53	13 1
Shortest passages........................	10 9 0	10 17 0	10 13
Bremen Line.—Average of 20 Eastern and 22 Western passages............	14 8 27	12 9 42	13 9
Shortest passages..........................	10 17 0	10 19 0	10 18

From the above it will be seen, that while the mean average of all the passages, made between Liverpool or Southampton and New York, ranges from 11 days up to 13 days 9 hours; it is estimated that by Ireland, Newfoundland, and Shippigan the passage could be made in 7 days 3 hours, nearly four days less time than the lowest mean average, and two days less than the shortest of 246 passages, if not the *very shortest* passage on record. These advantages alone are sufficient to attract the attention of business men, but the great recommendation of the Newfoundland route to most travellers, would be the shortening of the Ocean passage proper, from 264 hours (the average by the Cunard line) to 100 hours.

The above comparison has been made because the greatest number, and perhaps the best, Ocean Steamship Lines run to New York. A similar comparison with the Boston, Portland, and Quebec lines would show a result still more in favor of the Newfoundland route.

The following table, giving the time required between London and various points in North America, will show at a glance the great advantage which would accrue to the people of both hemispheres by the establishment of *the short Ocean passage Route.* By this table it will be seen that the Mails from London, could not only be carried to all parts of the British Provinces, and to all points in the Northern States, in a marvelously short space of time by the route herein projected, but that it is quite possible to deliver them on the shores of the Gulph of Mexico *in nine days,*—less time, in fact, than the shortest passages of the Cunard or of any other Steamers between Liverpool and New York.

Time required to carry the Mails by the Proposed Short Ocean Passage, and by the Intercolonial Railway from Shippigan.

From London to	St. Johns, N. F..................................	4 days	20 hours.
" "	Shippigan	5 "	20 "
" "	Halifax	6 "	5 "
" "	St. John, N. B.........................	6 "	4 "
" "	Quebec	6 "	10 "
" "	Montreal	6 "	16 "
" "	Toronto.................................	7 "	2 "
" "	Buffalo	7 "	6 "
" "	Detroit...................................	7 "	8 "
" "	Chicago	7 "	20 "
" "	Albany..................................	7 "	0 "
" "	New York..............................	7 "	3 "
" "	Boston	6 "	19 "
" "	Portland	6 "	15 "
" "	New Orleans..........................	9 "	0 "

Having shown that by shortening the ocean passage across the Atlantic to a *minimum*, the time of transit between the great centres of business in Europe and America can be very greatly reduced; so much so indeed that a reasonable hope may be entertained that the entire Mail matter passing between the two Continents, may eventually be attracted to the new route, it may be well now to enquire what proportion of Passengers may be expected to travel over it.

Before 1838 the only mode of crossing the Atlantic was by sailing ships: the passage commonly occupied from six to ten weeks, until the introduction of a superior class of vessels known as the American Liners; these fine ships made an average homeward passage of 24 days, and an average outward passage of 36 days.

The year 1838 saw the begining of a New Era in transatlantic communications. Two Steam vessels crossed from shore to shore; one, "The Sirius" left Cork on April 4th, another "The Great Western" left Bristol on April 8th, and they both arrived at New York on the same day, the 23rd of April; the average speed of the former was 161 miles per day, that of the latter 208 miles per day.*

"The Great Western" continued to run from 1838 to 1844, making in all 84 passages; she ran the outward trip in an average time of 15¼ days, and the homeward trip in an average time of 13¾ days.

The Cunard Line commenced running in July 1840, with three steamers, "The Britannia," "The Acadia," and "The Caledonia," under a contract with the British Government to make monthly passages.

In 1846, under a new contract, the Cunard Company undertook to despatch a Mail Steamer once a fortnight from Liverpool to Halifax and Boston, and another Mail Steamer once a fortnight from Liverpool to New York. This service has been maintained with amazing regularity and increasing efficiency to the present day.

These were the pioneers of a system of Ocean Steam Navigation which has already done so much to increase the intercourse between the two Continents. By reducing the length and uncertainty of the voyage as well as the inconveniences, in many cases the miseries, which passengers had previously to endure, a vast deal of good has been accomplished.

The number and tonnage of Steamships engaged in carrying passengers and goods between the British Islands and North America has of late years increased with wonderful rapidity. In 1864 no less than *ten regular lines* of Ocean Steamers were employed in running either to New York or to Ports north of that City in the United States or in Canada. Of these ten lines, two were weekly and eight fortnightly, equivalent in all to six weekly lines; so that there were on an average six Steamships leaving each side weekly, or nearly one every day.

The total number of passengers carried by these various Steam lines during the past year was 135,317, and by far the largest number travelled during the Summer months.

It would not take a very large proportion of Passengers crossing in any one year to give employment to *a daily line of Steamers* on the short Ocean Passage route from St. John to Valentia or to Galway. A total number of 40,000 each way would give 200 passengers each trip, for seven months in the year.

It is obvious then that there is already abundance of Passenger traffic, if the purely passenger route under discussion, possesses sufficient attractions. To settle this point the advantages and disadvantages of the route must be fairly weighed.

The obstructions offered by floating ice during several months in the year, are insuperable while they last; during this period Halifax or some equally good port, open in winter, will be available.

The frequent transhipments from Railway to Steamship, and *vice versâ*, may be considered by some an objection to the route; for conveyance of Freight they certainly would be objectionable, but most passengers would probably consider the transhipments, agreeable changes, as they would relieve the tedium of the journey.

With regard to the comparative safety of this route, it would seem as if the advantages were greatly in its favor. The portion of a voyage between New York and Liverpool, which seamen least fear, is that from Ireland to Newfoundland. It is well known that the most dangerous part of the whole voyage is along the American coast between New York and Cape Race, where thick fogs so frequently prevail; this coast line is about 1,000 miles in length and it has been the scene of the larger number of the disasters which have occurred. No less than fourteen or fifteen Ocean Steamships have been lost on this portion of the Atlantic Seaboard.†

*These are not claimed to be the very first Steamships that crossed the Atlantic, as, in 1833, five years earlier, a Canadian vessel "The Royal William" of 180 horse power and 100 tons burthen, sailed from Quebec to Pictou, N. S., and thence to London.

† *The following is a List of Ocean Steamships lost on the American Coast between New York and Cape Race. It may not be strictly correct, as it is compiled mainly from recollection:*

The Columbia.............................	on Seal Island, Nova Scotia.
The Humbolt	mouth of Halifax Harbour.

The route which favors increased security from sea-risks, and which is the shortest in point of time, must eventually become the cheapest and in consequence the most frequented. If then the route proposed across Newfoundland and Ireland avoids many of the dangers of existing routes and reduces the Ocean passage proper to 100 hours, would not the current of travel naturally seek this route in preference to others, especially when time would be saved thereby?

If, as it has been shewn, this route would reduce the time between London and New York some three or four days, and bring Toronto one third nearer Liverpool (in time) than New York is now; if it would give the merchant in Chicago his English letters four or five days earlier than he has ever yet received them; if it be possible by this proposed route to lift the Mails in London and lay them down in New Orleans in less time than they have ever yet reached New York, then it surely possesses advantages which must eventually establish it, not simply as an Inter-Colonial, but rather as an Inter-Continental line of communication.

These are purely commercial considerations, and however important they may be as such, the Statesman will readily perceive, in the project, advantages of another kind. It may be of some consequence to extend to Newfoundland, as well as to the other Provinces of British America, the benefits of rapid inter-communication. It will probably accord with Imperial policy to foster the Shipping of the Gulf and to encourage the building up of such a Fleet of swift Steamers as a Daily Line across the Ocean would require. It must surely be important to the Empire, to secure in perpetuity the control of the great Highway between the two Continents. It must be equally her policy to develope the resources and promote the prosperity of these Colonies—and to bind more closely, by ties of mutual benefit, the friendly relationship which happily exists between the people on both sides of the Atlantic.

The Chart which accompanies this will show, the important geographical position, which the British Islands and the British Provinces occupy, in relation to the shortest line of communication across the Ocean, between Europe and America.

The City of Philadelphia	Cape Race.
The Franklin	Long Island, New York.
The Indian	near Canso, Nova Scotia.
The Argo	near Cape Race.
The Hungarian	Cape Sable, Nova Scotia.
The Connaught	Bay of Fundy.
The Caledonia	Cape Cod.
The Anglo Saxon	Cape Race.
The Norwegian	St. Paul's Island, Atlantic side.
The Bohemian	Cape Elzabeth, Portland Harbour.
The Georgia	Sable Island.
The Pactolus	Bay of Fundy.

And another on Ragged Island, Nova Scotia, the name of which is not at present remembered by the writer.

www.ingramcontent.com/pod-product-compliance
Lightning Source LLC
Chambersburg PA
CBHW021946160426
43195CB00011B/1239